P9-BZP-013

ALSO BY

MARK BITTMAN

Mark Bittman's Kitchen Express

How to Cook Everything

How to Cook Everything Vegetarian

The Best Recipes in the World

Fish: The Complete Guide to Buying and Cooking

How to Cook Everything: The Basics

How to Cook Everything: Bittman Takes on America's Chefs

Mark Bittman's Quick and Easy Recipes
from *The New York Times*

WITH JEAN-GEORGES VONGERICHTEN

Simple to Spectacular

Jean-Georges: Cooking at Home with a Four-Star Chef

MARK BITTMAN

Food Matters

A GUIDE TO CONSCIOUS EATING WITH MORE THAN 75 RECIPES

Simon & Schuster Paperbacks

New York London Toronto Sydney

For Kerri Conan, Kelly Doe, Suzanne Lenzer,
and Sidney M. Baker, MD, who made this possible

This publication contains the opinions and ideas of its author. It is intended to provide helpful and informative material on the subjects addressed in the publication. It is sold with the understanding that the author and publisher are not engaged in rendering medical, health, psychological, or any other kind of personal professional services in the book. If the reader requires personal medical, health or other assistance or advice, a competent professional should be consulted.

The author and publisher specifically disclaim all responsibility for any liability, loss, or risk, personal or otherwise, that is incurred as a consequence, directly or indirectly, of the use and application of any of the contents of this book.

Before starting on a weight-loss plan, or beginning or modifying an exercise program, check with your physician to make sure that the changes are right for you.

Simon & Schuster Paperbacks
A Division of Simon & Schuster, Inc.
1230 Avenue of the Americas
New York, NY 10020

First Simon & Schuster trade paperback edition January 2010

Designed by Kelly Doe
Illustrated by Andy Martin

Manufactured in the United States of America

10 9 8 7 6 5 4

The Library of Congress has cataloged the hardcover edition as follows:
Bittman, Mark.
Food matters / Mark Bittman.
p. cm.
1. Nutrition. 2. Sustainable living. 3. Food habits. 4. Food industry and trade. 5. Agriculture. I. Title.
RA784.B55 2008
613.2—dc22 2008039593

ISBN 978-1-4165-7564-1
ISBN 978-1-4165-7565-8 (pbk)
ISBN 978-1-4165-7897-0 (ebook)

This book was printed using post-consumer recycled certified processed chlorine free paper and environmentally friendly inks.

Contents

Preface

IN THE EIGHTEEN MONTHS since I finished writing *Food Matters*, much has changed: food issues are at the forefront of national policy, and, with the (presumed at this writing) passage of a federally mandated plan for health insurance, the world of food will change even more, and even faster.

What hasn't changed is the logic behind this book: The way we eat in America makes us sick, and if we change our diet—simply, incrementally—we'll be healthier (and health insurance costs will decline). This doesn't mean, necessarily, that there is "bad" food. Eating habits can be "bad," or self-destructive (and, as you'll see, damaging to the planet). But food itself—real food, not junk—though it may be of varying quality, usually isn't "bad," in the sense that it will make you sick.

But unless you're blessed with fantastic and rare genes, the *way* in which you eat can certainly be quite harmful.

This distinction is subtle, and illustrated by the relatively recent demonizing of corn: There are people who now consider corn "bad." But corn is not the enemy; no single food is the enemy.

Could we be growing less corn? Absolutely. Could we be using the corn that we are growing more wisely? No question. And the fact that so many people are now addressing these and other policy issues is a great thing. You won't solve those problems by avoiding corn on the cob, tortillas, or even an occasional bowl of chips and salsa. But it might help if you stopped buying soda sweetened with high-fructose corn syrup every day.

In the time since *Food Matters* was first published, I've come to the conclusion that the food policy that matters most is yours. Your personal food policy is more potent than anything else you can do in the realm of climate change, animal welfare, the environment in general, and certainly your own health.

In fact, the way each of us eats has a powerful cumulative impact, and it was seeing that clearly that led me to write *Food Matters*. About three years ago, a report from the United Nations Food and Agriculture Organization (FAO) landed on my desk. Titled *Livestock's Long Shadow*, it revealed a stunning statistic: Global livestock production is responsible for about one-fifth of all greenhouse gases—more than transportation.

This was a signal moment for me, coming along with some personal health problems, an overall gloomy global outlook, and an increasing concern with animal products in general: the quality of meat, the endangerment of wild fish, the way domestic animals are raised, and the impact our diet has had on the environment.

Never before had I realized issues of personal and global health intersected so exquisitely. The destiny of the human race and that of the planet lay in our hands and in the choices—as individuals and as a society—that we made.

If I told you that a simple lifestyle choice could help you lose weight, reduce your risk of many long-term or chronic diseases, save you real money, and help stop global warming, I imagine you'd be intrigued. If I also told you that this change would be easier and more pleasant than any diet you've ever tried, would take less time and effort than your exercise routine, and would require no sacrifice, I would think you'd want to read more.

When you do, you'll find an explanation of the links among diet, health, the environment in general, and climate change in particular, and you'll learn how you can make a difference. And while you're doing your part to heal the planet you'll improve your health, lose weight, and even spend less at the checkout counter.

And yes: This is for real.

Introduction

TWO YEARS AGO, a report from the United Nations Food and Agriculture Organization (FAO) landed on my desk. Called *Livestock's Long Shadow,* it revealed a stunning statistic: global livestock production is responsible for about one-fifth of all greenhouse gases—more than transportation.

This was a signal moment for me, coming along with some personal health problems, an overall gloomy global outlook, and an increasing concern with animal products in general—the quality of meat, the endangerment of wild fish, the way domestic animals are raised, and the impact our diet has had on the environment. Never before had I realized issues of personal and global health intersected so exquisitely. The destiny of the human race and that of the planet lay in our hands and in the choices—as individuals and as a society—that we made.

If I told you that a simple lifestyle choice could help you lose weight, reduce your risk of many long-term or chronic diseases, save you real money, *and* help stop global warming, I imagine you'd be intrigued. If I also told you that this change would be easier and more pleasant than any diet you've ever tried, would take less time and effort than your exercise routine, and would require no sacrifice, I would think you'd want to read more.

When you do, you'll find an explanation of the links among

diet, health, the environment in general and climate change in particular and you'll see how *you* can make a difference. And while you're doing your part to heal the planet you'll improve your health, lose weight, and even spend less at the checkout counter. And yes: This is for real.

The consequences of modern agriculture

It doesn't take a historian to see that events that took place hundreds or even thousands of years ago reverberate to our day, and it doesn't take a scientist to see the profound effects of every significant advance in technology, from the invention of the wheel and the internal combustion engine to that of the microchip.

Unfortunately, we can rarely anticipate the consequences of historical events, inventions, and new technologies. Some have had nearly entirely positive results: indoor plumbing and vaccinations have saved countless lives, and it would be hard to argue that the telephone or railroads were not almost entirely positive. Automobiles, with their huge demand on limited energy sources, are a tougher call.

The industrialization of food production was one development that—though positive at first—is now exacting intolerable costs. Just as no one could foresee that cars would eventually suck the earth dry of oil and pollute the atmosphere to unsafe levels, no one could have anticipated that we would raise and eat more animals than we need to physically sustain us, that in the name of economy and efficiency we would raise them under especially cruel conditions (requiring some humans to work under cruel conditions as well), or that these practices would make them less nutritious than their wild or more naturally raised counterparts *and* cause enormous damage to the earth, including the significant acceleration of global warming.

Yet that's exactly what has happened. Industrialized meat production has contributed to climate change and stimulated a fundamental change in our diets that has contributed to our being overweight, even obese, and more susceptible to diabetes, heart disease, stroke, and perhaps even cancer.

It isn't just our propensity for eating animal products that's making us fat and sick, but also our consumption of junk food and overrefined carbohydrates. And these foods—which as a group are also outrageously expensive, especially considering their nutritional profiles—are also big contributors to environmental damage and climate change.

The twentieth-century American diet, high in meat, refined carbohydrates, and junk food, is driven by a destructive form of food production. The fallout from this combination, and the way we deal with it are issues as important as any humanity has faced: The path we take from this crossroads will determine not only individual life expectancy and the quality of life for many of us, but whether if we were able to see the earth a century from now we would recognize it.

Climate change is no longer a theory, and humans will suffer mightily if it isn't reversed. Most people know this. Less well known is the role that raising livestock plays in this, which is greater than that of transportation. Equally certain is that many lifestyle syndromes and diseases are the direct or indirect result of eating too many animal products. Our demand for meat and dairy—not our need, our want—causes us to consume way more calories, protein, and fat than are good for us.

Why food matters

Global warming, of course, was accidental. Even 30 years ago we couldn't know that pollution was more than stinky air. We thought it caused bad visibility and perhaps a few lung diseases here and there—as if that weren't bad enough.

The current health crisis is also an accident: We thought that the more meat and dairy and fish and poultry we ate, the healthier we would be.

This has not proved to be the case. Overconsumption has been supported and encouraged by Big Oil and Big Food—the industrial meat and junk food complex—in cahoots with the federal government and even the media and (one might say so-called) health industries. This has come at the expense of lifestyles that would have encouraged more intelligent use of resources—not just oil, but land and animals—as well as global health and longer life for individuals.

It doesn't have to continue: by simply changing what we eat we can have an immediate impact on our own health and a very real effect on global warming—*and* the environment, *and* animal cruelty, *and* food prices.

That's the guiding principle behind *Food Matters*, and it's really very simple: eat less meat and junk food, eat more vegetables and whole grains. I'm not talking about a diet in the conventional sense—something you do for two weeks or three months and then "maintain." I'm not suggesting that you become a vegetarian or eat only organic food. I'm not even talking about a method for weight loss, per se, though almost anyone who makes the kinds of changes I'm suggesting here is likely to lose weight and keep it off. You won't be buying exotic foods or shopping in expensive specialty markets, and you won't be counting calories—or anything else.

I'm just suggesting eating less of some things and more of others. The results will make you healthier while you do a little toward slowing climate change—much like trading in your gas guzzler for something more energy and cost efficient.

You could stop reading now and put your own plan into action. Or you can read on and find the details of how we allowed ourselves to be stuck with this mess and how you can help yourself and the rest of us get out of it. I'll describe what sane, conscious eating is, and the impact it will have. I'll suggest different strategies for changing how you think about food and prepare it. I'll show you how easy it is to follow the Food Matters plan when you eat out, whether at restaurants or other people's houses. I'll give you some sample menus and direction so you can easily create your own. Finally, I'm providing 77 easy recipes to get you started.

At first my suggestions may seem radical, but they can be integrated gradually into *any* style of eating. There's no sacrifice here, only adjustment and benefit: I will not suggest that you cut your calorie consumption (I don't even advocate counting calories), though you probably will simply by following the plan. Other than suggesting that you pretty much rule out junk food, I won't put any foods off limits.

The fact is that what I'm asking you to do isn't radical at all, and I'm confident you'll find this new mind-set so easy and so natural, and that you'll see its many benefits so easily, that you'll be eager to adjust your diet.

Why me?

Who am I to tell you how to eat and suggest it's a way to reduce global warming? I've been a reporter and researcher for more than 30 years; for much of that time, I've written about food from every possible angle. I've seen nutritional "wisdom" turned on its head more than once, and I've seen studies contesting studies designed to disprove studies. I have no more agenda than to inject some common sense into the discussion.

It doesn't take a genius to see that an ever-growing population cannot continue to devote limited resources to produce ever-increasing amounts of meat, which takes roughly 10 times more energy to produce than plants. Nor can you possibly be "nice" to animals, or respectful of them, when you're raising and killing them by the billions.

And it doesn't take a scientist, either, to know that a handful of peanuts is better for you than a Snickers bar, that food left closer to its natural state is more nutritious than food that has been refined to within an inch of its life, and that eating unprecedented quantities of animals who have been drugged and generally mistreated their entire lives isn't good for you.

I've got plenty of evidence to back up what I'm saying in these pages, but I've got my own story, too, and maybe you'll find that equally convincing. (It begins on page 71.) I've tried to strike a balance here, avoiding citing an overwhelming number of studies

Which Would You Choose?

Nutrient	2-ounce Snickers bar	2 tablespoons dry-roasted peanuts
Calories	271	107
Sugar	29 g	<1 g
Fat (saturated)	14 g (5.2 g)	9 g (1.8 g)
Protein	4 g	4 g
Sodium	140 mg	2 mg
Fiber	1 g	<2 g
Vitamin E	<1 mg	2 mg
Folate	12 mcg	27 mcg
Niacin	2 mg	3 mg

in an attempt to prove my point; that approach doesn't work, anyway, because most data can be read many ways, depending on your prejudices. My contention is that this way of eating is so simple, logical, and sane that cherry-picking scientific research isn't necessary.

One more thing: I'm not a doctor or a scientist, but I'm not a health-food or nutrition nut either. For my entire adult life I've been what used to be called a gourmand and is now called (unfortunately) a foodie: a daily and decent cook, a traveler who's eaten all over the world and written about it, a journalist and food lover who's eagerly devoured everything. I intend to continue to do just that, but in different proportions.

For our own sakes as well as for the sake of the earth, we need to change the way we eat. But we can continue to eat well—better, in fact. In the long run, we can make food more important, not less, and save ourselves and our planet (and some money) by doing so.

FOOD MATTERS PART I

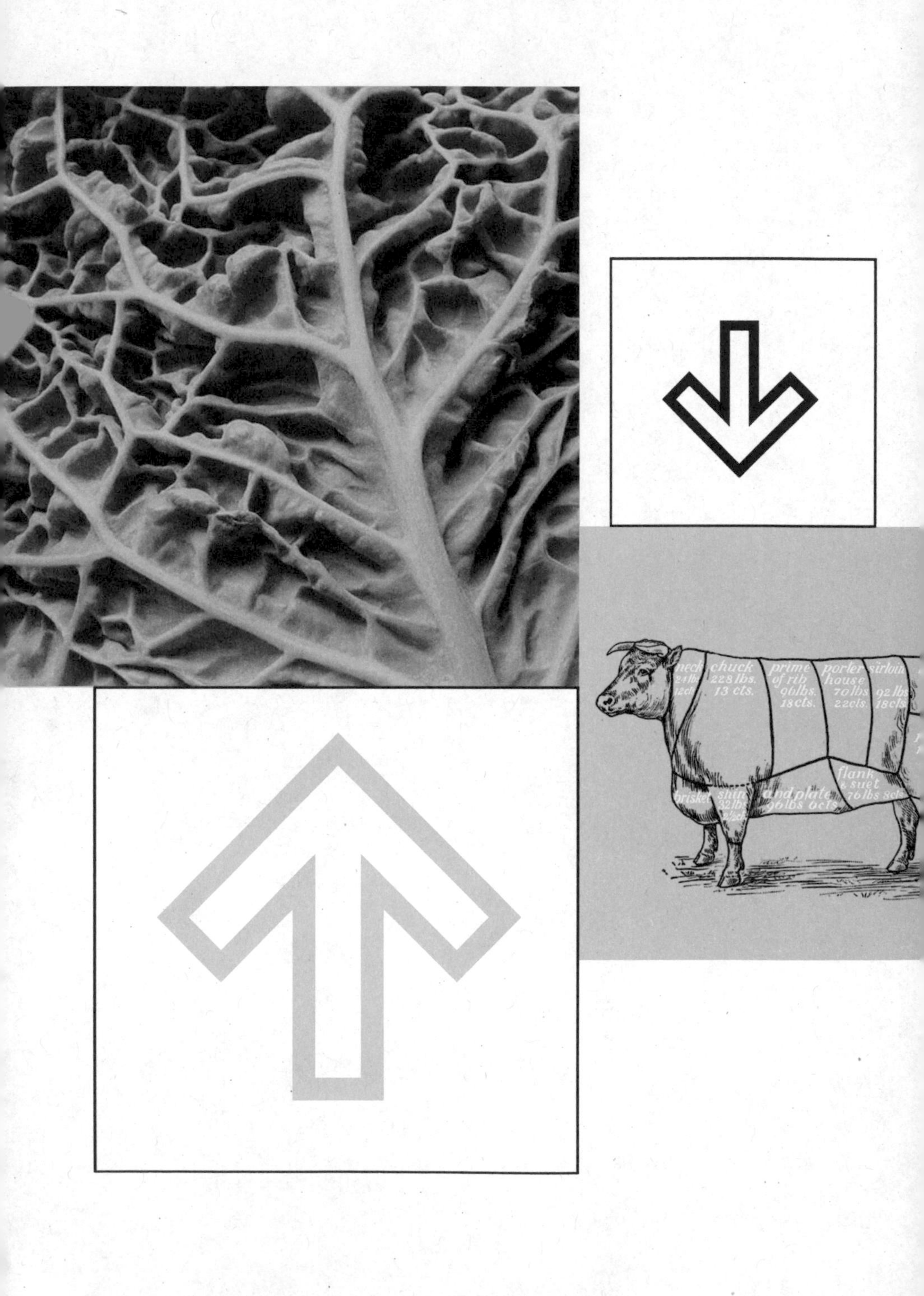
neck
24 lbs
12 cts
chuck
228 lbs.
13 cts.
prime
of rib
96 lbs.
18 cts.
porter
house
70 lbs
22 cts.
92 lbs
18 cts
flank
& suet
76 lbs 8 cts
brisket
shin
32 lbs
and plate
96 lbs 6 cts

Rethinking Consumption

Could improved health for people and planet be as simple as eating fewer animals, and less junk food and super-refined carbohydrates?

Yes. Of course health benefits for individuals would vary, and the effect on the planet would not necessarily be dramatic (as everyone knows, large adjustments in energy use are essential), but it would be a real step forward, and perhaps most important one that can be taken by individuals, with no government intervention.

There'd be other benefits, too: we would see the methods used in livestock production change. (This is important because the current system of raising animals for food is not only unsustainable but destructive, and will become more so: global meat consumption is expected to double within the next 40 years.)

The average person would also spend less money on food. With food prices in general rising at an average of about 5 percent a year, the differences in costs between vegetables, fruits, and grains, versus dairy, eggs, meat and especially junk food, are going to become more and more apparent (and painful, for those who refuse to make the change).

For the moment, let's ignore whether food is organic or local, or even whether animals are raised humanely. All these issues matter, but the bottom line here is that to eat well we must first eat moderately, and limit our eating to real food. (Organic junk food—and there is plenty of it—is still junk food.) Once we make those strides, which require small individual changes but whose collective impact is huge, we'll be able to eat more locally, we'll be able to eat more organic food, and we'll be able to treat animals more humanely. In fact, this will come naturally.

First, though, we have to adjust our consumption patterns. One argument, and it's a sound one, goes something like this: eat less meat, but eat better meat. "Better" meat, by its nature, tends to be local, more humanely raised, and less environmentally damaging: a good start. But my point, as I'll stress over and over, is that it all begins with eating *less* meat.

Our instincts, as human animals, prod us to eat all the food we can lay our hands on; difficult as it may be to imagine, until recently nearly all humans struggled to obtain enough calories. Those instincts, coupled with relative affluence, almost unlimited availability, and marketing that encourages us to eat the food that's most profitable for manufacturers, lead to overconsumption of precisely the wrong foods.

It's easy to see this with, say, fruits versus processed sweet snacks: It's far more profitable to produce and sell Twinkies and Cinnabons, for example, than to grow and sell strawberries. That's why so much more money is spent convincing us of the deliciousness of Twinkies and Cinnabons.

Similarly, it's more profitable to sell a million pounds of industrially raised meat than it is to sell 100,000 pounds of humanely raised, antibiotic- and hormone-free. And if you're the producer of that meat, you create demand as necessary. Maybe you lower prices. Or you tell consumers that meat is healthier than an alternative protein source. Or you make it more appealing: it's manly, it's real food, it's what's for dinner. Maybe you even cook it for them and sell it as cheaply as you can. Or you provide a combination of all of these, which is what we have today. Whatever it takes.

The people in many developed countries, including the U.S., consume 1/2 pound of meat per day.

Most people crave meat. Arguably, that craving is natural, or at least not unnatural. We are omnivorous, capable of digesting a wide range of foods, and historically we have eaten just about all of them, first from necessity and then for pleasure.

If you grow up eating meat *and* it's marketed as real, healthy, cheap, sexy, and delicious, you really enjoy eating it. But given a large enough marketing budget, we can be persuaded to eat just about anything, including concoctions that contain no naturally occurring food at all.

A new world of meat eaters

We might love meat, we might benefit from eating it in moderate quantities, but we don't *need* to eat meat to live. And most independent experts believe that consuming it at our current levels is bad for us. And our consumption is headed in the wrong direction. Livestock, globally, is the fastest-growing sector of agriculture: Since 1980 the global production of pigs and poultry has quadrupled, and there are twice as many cattle, sheep, and goats.

The people in many developed countries (including the United States) consume an average of about half a pound of meat per day; in Africa, the average is about an ounce a day. And though meat consumption is fairly stable in the United States, it's rising at a faster rate in the developing world, where it has tripled since 1970. The Chinese eat twice as much meat as they did a decade ago.

Between 1995 and 2005, the number of chickens world-

Past and Projected Food Consumption of Livestock Products

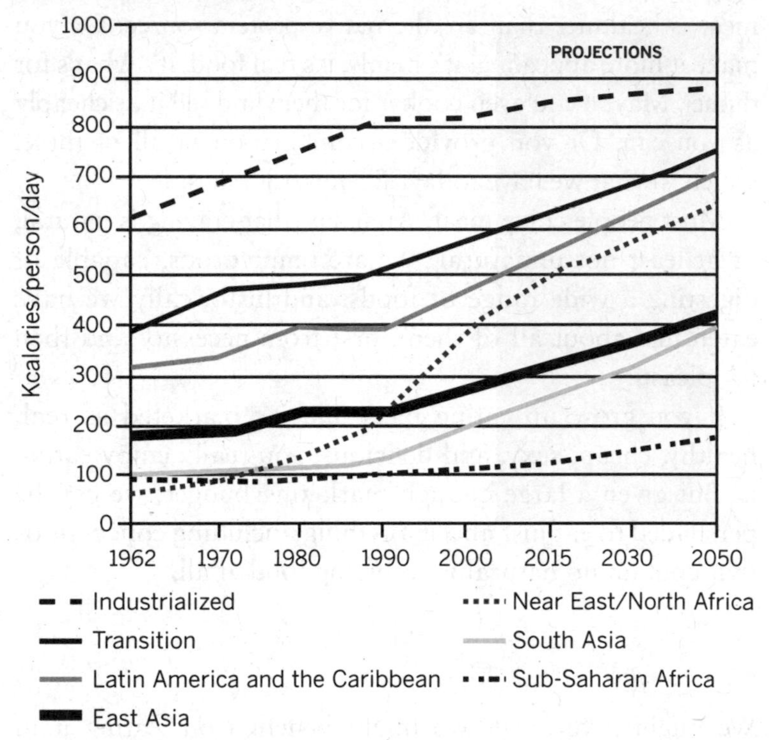

Note: For past, three-year averages centered on the indicated year. Livestock products include meats, eggs, milk, and dairy products (excluding butter).

Source: FAO (2006a) and FAO (2006b).

wide destined to be eaten rose by 14 billion (an increase of 40 percent); the number of egg-layers increased by 2.3 billion (31 percent); the number of pigs rose by 255 million (24 percent); and the number of cows used for milk production increased by 12 million (6 percent). The FAO predicts that this increase in animal production will continue, and that meat production will double again by 2050.

Impressive numbers. And the truth is that because of them, the world *needs* factory farming. There is no other method that can produce these quantities of meat, eggs, and dairy. It follows, then, that *the only way to reduce factory farming is to demand less meat.*

We currently raise 60 billion animals each year for food—ten animals for every human on earth. The projection is that just to sustain current consumption levels (and consumption is increasing, so this is conservative), by 2050 we'll be raising 120 billion animals a year.

60 BILLION animals are raised each year for food—10 animals for every human on earth.

That number would require using more land for agriculture than exists. Even if we could find the space (or technology) to meet the demand, the number also assumes that the atmosphere, land, and oceans could tolerate it. The effect would be cumulative, like credit card debt: a year of animal consumption at this rate requires a year and two months' worth of resources. And since consumption is increasing, the situation will get worse even faster. In developing parts of Asia, for example, meat consumption increased 131 percent between 1980 and 2002; in Latin America and the Caribbean, 24 percent; in industrialized countries, 10 percent; and in the world as a whole, 22 percent.

The only way to reduce factory farming is to demand less meat.

It's not just meat

There's another aspect to this problem, one that many experts believe affects our health even more dramatically than meat. And though it's been overshadowed by livestock in the realm of ecological damage, it's equally alarming.

That is the world of junk food, overrefined carbohydrates, and highly processed oils—foods that make up an astonishingly large part of our diet. A study from the University of California at Berkeley, for example, reports that almost one-third of Americans' total caloric intake comes from "nutrient-poor" foods like sweets, salty snacks, and fruit drinks. Seven percent of our calories come from soda—more than from vegetables—with hamburgers, pizza, pastries, and potato chips following close behind. (See the chart on the next page.) Meanwhile, beef, pork,

Top 10 Foods Contributing to Energy Intake in the U.S. Population

Rank	Food	% of Total Energy
1	Regular soft drinks	7.1
2	Cake, sweet rolls, doughnuts, pastries	3.6
3	Hamburgers, cheeseburgers, meat loaf	3.1
4	Pizza	3.1
5	Potato chips, corn chips, popcorn	2.9
6	Rice	2.7
7	Rolls, buns, English muffins, bagels	2.7
8	Cheese or cheese spread	2.6
9	Beer	2.6
10	French fries, fried potatoes	2.2

7% of Americans' calories come from soda.

dairy, chicken, and fish account for 23 percent of our total caloric consumption, while vegetables and fruit—including juice, which is often sugar-laden—barely hit 10 percent. (See the chart on the opposite page.)

The term "junk food" means different things to different people. Potato chips. Shakes. Candy. Doughnuts. Double cheeseburgers. Chicken nuggets. White bread. None of these has as justifiable a role in good eating as decently raised meat, poultry, or fish; but all of them (with the possible exception of candy) represent categories of food that, made well and eaten occasionally, have a traditional and even legitimate role. None is, by itself, "junk." But from our bodies' point of view, they all may do more harm than good. Why?

Top 10 Food Groups Contributing to Energy Intake in the U.S. Population

Rank	Food	% of Total Energy
1	Sweets, desserts	12.3
2	Beef, pork	10.1
3	Bread, rolls, crackers	8.7
4	Mixed dishes	8.2
5	Dairy	7.3
6	Soft drinks	7.1
7	Vegetables	6.1
8	Chicken, fish	5.7
9	Alcoholic beverages	4.4
10	Fruit, juice	3.9

For the most part, these foods contain far more calories than are justified by their nutrient levels. In part, this is because they're largely made from corn, in the form of a sugar called high fructose corn syrup; soy in the form of extracted protein or oil; or refined wheat—white flour—all processed to the point where they're nutritionally worthless or even damaging. (See The Force-Feeding of America, page 44.) Furthermore, they often contain added ingredients like preservatives and other chemicals that are at best useless and may be harmful. (Not coincidentally, corn, soy, and wheat are among our most highly subsidized and environmentally damaging crops. (See Nutrition Advice Meets Food Policy, page 43.)

Consider the difference between eating a whole baked potato and eating an individual bag of potato chips. You'd need to eat 2.5 ounces of potato chips (that's two and a half single-serving bags, less than what most people eat in a

Two Forms of Potato

Nutrient	Potato Chips (2.5 ounces)	Medium Baked Potato with Pat of Butter
Calories	380	204
Fat	25 g	4 g
Sugar	0 g	2 g
Protein	5 g	5 g
Fiber	4 g	4 g
Vitamin C	22 mg	22 mg
Folate	33 mcg	45 mcg
Potassium	903 mcg	953 mcg

sitting) to get the protein in one medium baked potato. By then you would have consumed nearly 25 grams of fat and 380 calories; nearly twice the amount in the baked potato, *even with a pat of butter*.

The environmental impact of overconsumption

Even the most conscientious agriculture has some environmental impact, and though much food production yields greenhouse gases, raising livestock has a much higher potential for global warming than crop farming. For example: To produce one calorie of corn takes 2.2 calories of fossil fuel. For beef the number is 40: *it requires 40 calories to produce one calorie of beef protein.*

In other words, if you grow corn and eat it, you expend 2.2 calories of energy in order to eat one of protein. But if you process that corn, and feed it to a steer, and take into account all the other needs that steer has through its lifetime—land use, chemical fertilizers (largely petroleum-based), pesticides, machinery, transport, drugs, water, and so on—you're responsible for 40 calories of energy to get that same calorie of protein. According to one estimate, a typical steer consumes the equivalent of 135 gallons of gasoline in his lifetime, enough for even some gas guzzlers to drive more than halfway from New York to Los Angeles, or for an energy-efficient car to make the drive back and forth twice. Or try to imagine each cow on the planet consuming almost seven barrels of crude oil.

40
CALORIES
of fossil fuel are required to produce 1 calorie of beef protein.

Another way to put it is that eating a typical family-of-four steak dinner is the rough equivalent, energy-wise, of driving around in an SUV for three hours while leaving all the lights on at home. In all, the average American meat eater is responsible for one and a half tons more CO_2-equivalent greenhouse gas—enough to fill a large house—than someone who eats no meat. If we each ate the equivalent of three fewer cheeseburgers a week, we'd cancel out the effects of all the SUVs in the country. Not bad.

Yet thanks to agricultural subsidies and the lack of regulation about how meat is raised, it's far less expensive than it actually should be.

Because it's more difficult to get at the raw data, it's not as easy (or as much fun) to make similar statements about junk food. But when you add in all the packaging required to get the stuff into supermarkets and fast-food restaurants, the environmental damage is impressive enough. One estimate is that the food industry accounts for 10 percent of all fossil fuel used in the United States; of this, the total energy expended by processing, packaging, and transportation of food products is 37 percent.

To give you an idea of how much more energy goes into junk food than comes out, consider that a 12-ounce can of diet soda—containing just 1 calorie—requires 2,200 calories to produce, about 70 percent of which is in production of the aluminum can. Almost as impressive is that it takes more than 1,600 calories to produce a 16-ounce glass jar, and more than 2,100 to produce a half-gallon plastic milk container. As for your bottled water? A 1-quart polyethylene bottle requires more than 2,400 calories to produce.

Overproduction drives overconsumption, which in turn is bad for our bodies and the environment—but these negative effects can be diminished by more moderate consumption, which in turn will eventually lead to lower production. This is where we come in: Every time you drink a glass of tap water instead of bottled water, you save the calorie equivalent of a day's food: the 2,400 calories it takes to produce that plastic bottle.

Likewise, every time you eat a salad instead of a burger you save energy. Look at it this way: When you eat a quarter pound of beef, you're consuming about 20 percent of your daily calories, but it takes about 1,000 calories—almost half your daily intake—to produce that burger. Remember, beef production requires energy for processing, transportation, marketing, and, most of all, the production of all the grain fed to the cow in the first place. (Producing a salad requires energy too, but nothing like what it takes to make that quarter-pounder.) Whenever you eat what might be called inefficient food—and beef is among the leaders in this category—you're consuming more of the planet's energy than you need to live well.

2,200 CALORIES are required to produce a 12 OZ. can of diet soda.

To make the case for changing your diet even more compelling, consider this: For a family that usually drives a car 12,000 miles a year, switching from eating red meat and dairy to chicken, fish, and eggs just one day a week—in terms of greenhouse gas emissions—is the equivalent of driving 760 miles less a year. And if you switch to a vegetable-based diet for that one day a week, you reduce emissions even more, to the equivalent of driving 1,160 miles less.

And this impact is exponential: By moving totally away from red meat and dairy to a diet made up of chicken, fish, and eggs you reduce your emissions by a further 5,340 miles a year. And if you switch to a completely vegetable-based diet? That same family reduces its emissions by more than 60 percent; the same as cutting their mileage down from 12,000 to just 3,900 miles a year.

Meat consumption would have to fall

3 OZ. A DAY

to stabilize greenhouse gasses produced by livestock.

Those are simple steps. But if, as is expected, the global population grows by nearly half in the next 40 years, meat consumption would have to fall to about three ounces a day (less than half of what Americans average now) just to stabilize the amount of greenhouse gases produced by raising livestock.

And stabilizing production isn't going to cut it, since even at current levels global warming is deadly. But since our consumption of energy would also have to be cut back, let's take this as a goal.

The choice is obvious: To reduce our impact on the environment, we should depend on foods that require little or no processing, packaging, or transportation, and those that efficiently convert the energy required to raise them into nutritional calories to sustain human beings. And as you might have guessed, that means we should be increasing our reliance on whole foods, mostly plants.

But before we move forward, let's take a look back to how we got to this place.

A Brief History of Overconsumption

Everything I've discussed so far—the overproduction and consumption of meat, the omnipresence of junk food, our declining health, the contribution of agribusiness to global warming and other environmental horrors—happened gradually: A hundred years ago, none of this was in sight. But though it began slowly, the process accelerated wildly following World War II, and went out of control 20 or 30 years later.

Meat's industrial revolution

How did meat production became industrialized? How did the family farm become the factory farm?

In 1900, 41 percent of American workers were employed in agriculture; now, that number is less than 2 percent. Many of us still live in or near rural areas, though, and even city dwellers can sense the familiarity, obligation, affection, and gratitude traditional farming families must have felt. There's no time off from farm work but the payoff—whether

Animals killed each year in the U.S. for food:

9 billion CHICKENS

100 million PIGS

250 million TURKEYS

36 million COWS

eggs, milk, meat, or all of these, plus hides, leather, fertilizer, pillows, and more (companionship, too, of course)—made raising animals a natural part of life.

Until the early twentieth century most animals were treated in much the same way they had been for a couple of thousand years. Raising more animals than your family could use was always a way to augment the family income; but it was to feed an increasingly urban population in the twentieth century that farmers starting raising chickens for meat as well as eggs, and moved cattle and pigs into feedlots, the progenitors of the modern confined and feeding operations (CAFOs).

As should be expected in a society with few limits on business, there was an opportunity to make real money on food animals. Since they were destined for death anyway, it made sense—from a purely economic perspective, at least—to raise them as efficiently as possible. For better or worse, the human mind is malleable enough to consider raising animals destined for the table not much differently from making plastic, even while keeping pets in the house.

Perhaps no one could have seen the result. But today's factory farm is a living hell that has far more in common with factories—places where things are mass-produced in the quickest and most cost-efficient manner possible—than it does with people working the land and raising animals. This kind of farming is accelerating globally, but it's a mature industry in the United States, where nearly all of our food requires some form of mechanization, synthetic chemicals, drugs, refrigeration, heating, cooking, radiation, freezing, long-distance transportation—or a combination of any or all of these.

The number of animals killed in the United States each year is staggering, something like 9 billion chickens, 36 million cows (including 1 million for veal), 100 million pigs, and 250 million turkeys. These numbers swell further when you consider the dairy cows (9 million) and egg-laying chickens (300 million), which aren't intentionally put to death but live in conditions that most Americans would consider unbelievably cruel if they were applied to dogs, cats, parakeets, or any other animal not customarily eaten.

Cheap soy and cheap corn yield cheap meat (and cheap lives)

There are, at first glance, advantages to all this, or at least one advantage: even with rapidly rising costs, meat remains relatively inexpensive. At an average of $1.69 per pound for chicken, $2.85 for pork, and $4.11 for beef, with double cheeseburgers still going for 99 cents and meat-based "casual dining" meals at around $10, the vast majority of Americans can easily afford to eat meat at least once a day, and often more.

But neither factory farming nor our junk food habit could exist without cheap corn and soybeans (wheat plays a role, too, but a slightly lesser one).

More than 50% of the corn grown in the United States is fed to animals.

There's nothing intrinsically wrong with corn or soy; whole cultures have relied on each as their main source of nourishment. But in the United States, and increasingly around the world, an overwhelming proportion of farmland is devoted to growing these two crops, not for us to eat directly (the most commonly grown varieties are not fit for human consumption), but to feed to animals or convert to oil or sugar. So dominant have these crops become (wheat, rice, and cotton are the other giants), that America no longer grows enough edible fruits and vegetables for everyone to eat our own government's recommended five servings a day. Were we all to do so, we'd be dependent on imported vegetables!

More than 50 percent of the corn grown in this country is being fed to animals; of the remainder, most finds its way into junk food (usually in the form of high fructose corn syrup), corn oil, and ethanol.

1
billion
people in the world are chronically hungry;
1
billion are overweight.

The story of soy is similarly dismal: Nearly 60 percent finds its way into processed food; the rest is used to make soy oil and animal feed (globally, 90 percent of soy meal is fed to animals). This makes it easy to understand why more than 1 billion people around the world are overweight. (The trendy term for this phenomenon is "globesity.")

Even more distressing is the sheer waste of feeding corn and soy to animals and using it to produce junk food. There are nearly a billion chronically hungry people on our planet, and we have the means—the food, even—to nourish them. According to the FAO, "world agriculture produces 17 percent more calories *per person* today than it did 30 years ago, despite a 70 percent population increase" (emphasis added). The researchers estimate that's about 2,720 calories per person, per day. To help visualize this absurdity, consider that the beef in one Big Mac is equivalent—in terms of grain produced and consumed—to five loaves of bread. But instead of feeding the hungry with grain, a lot of it is going to the waistline of people in wealthy countries—often to their detriment.

It's no exaggeration to say that soy and corn are killers, whether directly (soy oil is used to make trans fat, and high fructose corn syrup is about the most useless form of calories ever created) or indirectly (their cultivation is an environmental nightmare, and as animal feed in factories they're perpetuating a destructive system). And their use in these capacities is depriving millions of the food they desperately need. If we simply shifted resources to growing crops that fed people directly, we'd go a long way toward resolving many issues of health, agriculture, and the environment.

Soy, corn, and American farmland

Traditional farming—regenerative farming, as it's sometimes called now—relies on crop rotation to rest and replenish the soil. This method goes back thousands of years, and is mentioned in the Bible, which mandates that fields remain fallow one year in seven, to give them time to recover.

Different crops use different nutrients, so that planting certain crops sequentially may actually improve soil quality. "Cover" crops, another part of this plan, are grown primarily to return nutrients to the soil. With care, this kind of farming can be done productively and organically, without the aid of man-made chemicals.

Planting only one or two crops is called monoculture, and this type of farming is typically used for commodity crops, like corn (which accounts for 27 percent of the harvested crops in the United States) and soy (another 25 percent). Monoculture doesn't return nutrients to the soil, so it can't be effective without the aid of chemical fertilizers, which in turn consume huge resources of energy because they're based on fossil fuels that must be refined and usually transported long distances.

Together, soy and corn account for 50% of the total U.S. harvest.

Plus, chemicals do nothing to replenish micronutrients or the beneficial characteristics of the land. Our soil, once this country's most valuable resource, is not only becoming depleted, it's literally vanishing. In other countries, so is the forest: Demand for soy, primarily for animal feed, is a principal agent in deforestation in South America.

So, you have two crops coming to dominate first U.S. farmland and then global farmland. You have forests destroyed to grow more of these crops. As a result, you have diminishing resources. These two crops produce food that is either next to useless or damaging to humans when consumed in large quantities. When it is fed to animals, it is inefficient (remember, up to 40 times as much energy is needed to produce one calorie of meat as to produce one calorie of grain) and environmentally destructive.

Furthermore, cows were never meant to eat soy or corn; the digestive system of cows developed to eat grasses. But you cannot possibly raise as many cattle as are sold on pasture, or as many pigs in sties, or as many chickens in

50%
of the antibiotics administered in the United States go to animals.

yards, so producers had to figure out another, more "efficient" way to raise these animals. That way is confinement: sometimes in pastures, sometimes in cages, sometimes in concrete, almost always with soy and corn as feed. (It's actually even worse: Although chickens, pigs, and cows are herbivores, naturally foraging for plant food, we've turned them into carnivores, often supplementing their grain with ground-up animal parts.)

The combination of crowded living conditions and unnatural feed makes the animals vulnerable to disease, so they're often given subtherapeutic antibiotic treatment to keep them just healthy enough to survive, put on weight, and get to market—fast. (Half of the antibiotics administered in the United States go to animals, not humans; cattle are also routinely given growth hormones.) Feeding animals antibiotics increases antibiotic resistance in humans, and though the Food and Drug Administration (FDA) says hormones do humans no harm, other people believe the jury is still out on this one.

In fact, unless you're one of the people making millions or billions of dollars from this system, it's all bad.

Why we can't (yet) be nice to animals

I'm going to skip most of the deplorable stuff about factory farming. In fact any detailed description of growing animals industrially (the word "raising" is really misapplied here) would sicken anyone who has even the slightest feeling for other species (this includes all pet owners who are not extreme hypocrites), or who believes that the earth is to be shared by all creatures (except maybe mosquitoes), or who believes in fairness, justice, or kindness.

Let's cut to the chase. From conception to death, everything about the living conditions of animals raised in confinement is horrific. And the food these animals produce is unnatural, drug-tainted, and—compared with food from traditionally raised animals—tasteless. Furthermore, the impact on the environment and our bodies is, if not yet catastrophic, then certainly damaging.

Yet as much as we may despise this situation, we cannot resolve the plight of our fellow animals simply by deciding to be kind. We may be able to improve their existence *slightly*, for example by mandating an increase in the space allotted for each individual animal (though it's hard to imagine any government agency doing so). But even this wouldn't help much, because such changes cannot possibly be significant at our current levels of consumption.

At first look, grass-fed beef is an attractive alternative, since pastured cattle are in a more natural environment than those raised in confinement. But regardless of whether grass-fed beef is better for the health of the planet—and it's not entirely clear that it is—the world's pastures cannot even come close to supporting the 1.3 billion cattle currently being produced.

About 70% of the world's farmland is dedicated to livestock production.

And as demand for meat is increasing, any additional grass-fed cattle would not replace those raised in confinement but just add to their numbers. Since about 70 percent of the world's farmland—one-third of the ice-free land surface of the earth—is dedicated to livestock, directly or indirectly (by indirectly I mean, for example, growing feed), to raise the amount of beef on grass that is currently being produced in confinement would mean destroying nearly all existing forests and farmlands.

There isn't enough room. Factory farming was developed in large part to consolidate resources and make it possible to raise enough animals to meet inflated demand. If you hate factory farming (and you should), your primary concern should be reducing consumption.

Even if we *could* treat animals more humanely and maintain current production numbers—logically impossible, but let's pretend—if we were to give the animals more space, better food, fewer drugs, access to outside, and so on—the environment would still suffer, to their detriment and ours.

Production will not decrease as long as it's profitable, so we need to reduce it, and we can do so only by reducing demand. (Production would be less profitable with stricter laws and law enforcement, and with lower subsidies for corn and soy, but the federal government, no matter who's running it, shows no inclination to move in that direction.)

It's much like energy: Successive governments have encouraged and supported an oil-based economy while discouraging more sustainable forms of energy, knowing all the while that the result would be pollution, war, and rising costs. And still, it's clear there isn't enough oil for all.

As with oil, our governments have supported the production of meat and the transformation of grain into refined carbohydrates, knowing that high consumption of both would use farmland that could be better used in other ways, that it would damage the health of individuals and, now, that it has contributed to global warming. And still there isn't enough for all.

Production will not decrease so long as it's profitable—we need to reduce demand.

Even if the America-first argument were pragmatic, no one with a conscience could seriously argue that Americans are entitled to eat more meat than people in other countries, just as no one can argue that only we should have the freedom to drive cars. And just as we must reduce our consumption of energy—and we will, either by planning to do so, by increased demand for limited resources by developing countries, by seeing a worldwide energy crisis, or most likely by all three—we must reduce our consumption of meat and dairy food. In fact, if the developing world increases its meat consumption to a level approaching ours, that would amount to committing global suicide.

Until just a couple of years ago, when people talked about the effects of industrialized meat production on the environment, they were talking about rampant antibiotic usage; contamination of local land and water by fertilizers used to produce feed; the impact of pesticides and herbicides; the devastation of the world's forests, cleared for land on which to raise more livestock, or for their feed; the stink created by (in particular) pigs grown in confinement, and the effect this has on human (and other) neighbors; water usage; and a host of less well-publicized issues.

Livestock produce more greenhouse gas than the emissions caused by transportation.

These are all important, but the creation of greenhouse gas trumps them all. Livestock produce more greenhouse gas than the emissions caused by transportation or anything else except energy production. Add to this the humane and human health issues and one could easily and sensibly argue that it makes more sense to cut down on eating meat than it does to cut down on driving.

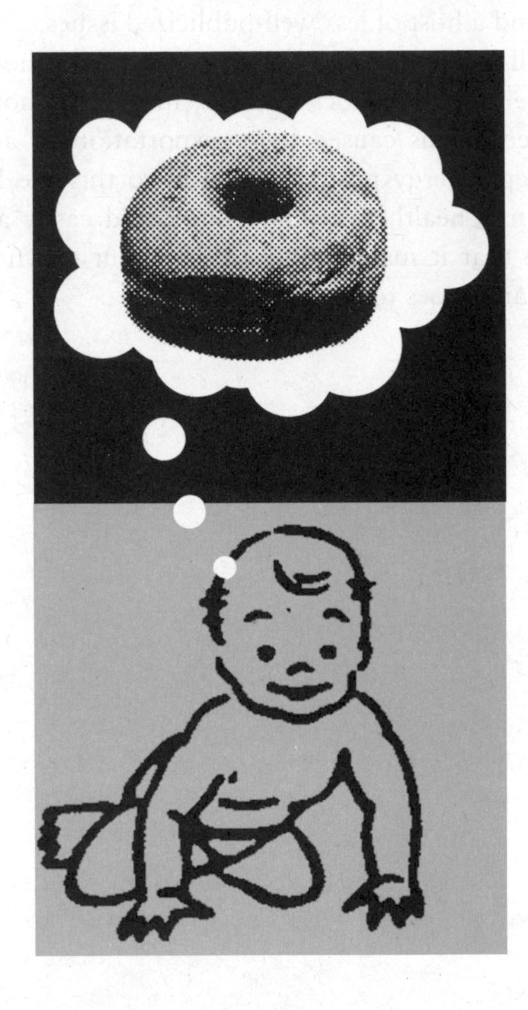

Selling the Bounty

Factory farming, the overproduction of corn and soy, junk food—these are just the most obvious examples of an agricultural production and marketing system gone awry: one that promotes consumption of foods that are detrimental to health, rather than real and wholesome foods.

Food companies, agricultural marketing boards like the National Chicken Council, and huge public relations and advertising firms (what was once loosely referred to as Madison Avenue) form a powerful alliance—Big Food meets Big Ad—promoting consumption of their products in obvious ways like advertisements and branding, and in much more devious, subtle ways, like funding research and paying supermarkets for shelf space.

That Big Food uses all the marketing tools at its disposal is no surprise; so does every other profitable producer, from Exxon Mobil to Pfizer. In the food world, as elsewhere, advertisements, licensing arrangements, and promotional campaigns are omnipresent. They're successful, too: A 10-year deal between McDonald's and Disney was worth about $1 billion; a similar alliance between Pepsi and the *Star Wars* movies was estimated to be worth $2 billion.

The Web has made promotion easier than ever, offering

infomercials in the form of recipes (usually a little scary; for instance, look at www.kraftfoods.com), so-called health updates (often questionable, like the yogurt promoted on www.activia.us.com, with its two-week challenge to improve "intestinal transit"), and online communities (always blatantly self-serving, like McDonald's African-American "community" site, www.365black.com).

Advertising in disguise

No one, of course, is forced to look at this junk. But it affects not only children (who, though many people insist they be "protected," often become true believers), but obviously adults as well, most of whom are ill-equipped to fend off the greatest marketing assault in history.

It's the promotion that flies under the radar that is really insidious. Few Americans know that almost all foods grown in the United States are promoted through nonprofit marketing organizations known as commodity boards, which are ostensibly regulated by the U.S. Department of Agriculture (USDA)—a perfect example of the fox guarding the chicken coop.

Most of us are ill-equipped to fend off the greatest marketing assault in history.

The idea is that all producers contribute proportionally to a fund, usually called a checkoff. This money finances research, public "education," lobbying, advertisements and other promotions to increase consumption of the food in question in the United States and abroad. (The boards also develop standards of labeling, handling, and safety, but by far the largest amount of their revenue goes to promotion.) You've seen these boards' work everywhere, probably without knowing it: "Got Milk?" "Pork—The Other White Meat," "The Incredible Edible Egg," and "Beef: It's What's for Dinner" are among the best-known examples, and among the most successful marketing campaigns in history.

Almost every food you can think of, from avocados to walnuts, has a marketing board. The more sales a product generates (beef, $71 billion per year; avocados, $489 million per year), the bigger the marketing program. The goal, as always, is to increase sales by whatever means necessary, and these means include huge budgets for nonstop lobbying to affect policy and legislation. The lobbying arms, usually called government affairs committees, can have a powerful effect on legislation. (It helps when industry members become government officials, too, in a typical revolving-door policy that demonstrates the power industry has over and in government.)

Food manufacturers add or change ingredients to make their products appear healthier.

Just like food corporations, food boards use every means at their disposal to convince us of the "health" of their products and the benefits of consuming ever more of them, even when there's evidence to the contrary. Nearly every move the boards make is aimed at skewing public perception, with the goal of increasing sales.

A quick look at www.gotmilk.com demonstrates this. Want to "rebuild muscle" or "increase stamina"? You don't need soy protein or Gatorade (or, god forbid, water); milk or chocolate milk is what you should reach for. Want an excuse to drink more coffee, or eat more cookies (specifically, in one recent campaign, Oreos)? Add milk. Want to make sure your daughters don't suffer from loss of bone density later in life? Make sure they eat a "healthy" breakfast—with plenty of dairy food. Want to sleep better? Drink milk before bedtime.

Just about every food board stresses the results of studies (often funded by the board itself; see page 63) showing that in one way or another, their food is "good and good for you"—even when it isn't.

Food boards stress studies that show their food is "good and good for you"—even when it isn't.

Food manufacturers take a similar route. They add or change ingredients to make their products appear healthier; they'll sneak their "healthy" product into otherwise less desirable foods; they make health claims that are borderline nonsense.

In short, they'll do whatever it takes to sell the product.

Yogurt and Ice Cream—What's the Difference?

Nutrient	Plain Whole-Milk Yogurt (1 cup)	Low-Fat Vanilla Yogurt (1 cup)	Low-Fat Vanilla Ice Cream (1 cup)	Vanilla Ice Cream (1 cup)	"Vanilla Truffle" Whole-Milk Yogurt (1 cup)
Calories	149	208	251	265	293
Sugar	11 g	34 g	34 g	28 g	48 g
Fat (saturated)	8 g (5 g)	3 g (2 g)	7 g (4 g)	15 g (9 g)	7 g (4 g)
Protein	9 g	12 g	7 g	5 g	8 g
Calcium	296 mg	419 mg	245 mg	169 mg	335 mg (est)
Potassium	380 mg	537 mg	316 mg	263 mg	346 mg

Take, for example yogurt, long considered a healthful food containing beneficial bacteria. Most yogurt is now flavored and sweetened so that it's more akin to ice cream than to regular milk, and the beneficial live cultures are often killed during processing.

Yogurt has been added to cereals, snack bars, fast-food-chain menus, makeup, and toothpaste. It's been altered to emphasize its "health" benefits: DHA and omega-3 have been added to some brands, and Dannon's Activia contains a proprietary bacteria that the company claims is "clinically proven to help regulate the digestive system."

All this works: Yogurt consumption has more than doubled in the last 20 years.

Then there's cereal. Once upon a time, cereal meant oatmeal, or even relatively benign corn flakes or puffed rice. But many of today's packaged cereals have long lists of ingredients. You may see a brand that touts itself as "multigrain," which means nothing more than having "more than one flour." Though the word implies some health benefit, it can be as simple as putting oat flour into a cereal that is effectively a boxful of small cookies—this can hardly have an impact when a cereal contains 30 percent sugar, as many do. Yet a claim like "may help reduce the risk of heart disease" may be featured on cereals that contain less than a gram of fiber, even though many experts believe that 25 grams a day is a desirable target—so it's like adding vitamin C to a candy bar and maintaining that the candy is good for you.

The marketing of cereal and many other foods is often shamelessly aimed at children. Like Big Tobacco, Big Food is relentless in its quest for an impressionable, long-term base, and children are the perfect target. The strategy works, too: in 2007, a couple of much-touted studies showed that when children were offered the same food in two packages—one labeled McDonald's and one unbranded—they preferred McDonald's. Similarly, children preferred the taste of carrots and milk if they thought these were from McDonald's.

Children preferred the taste of carrots and milk if they thought these were from McDonald's.

The marketers have invaded every possible space. Food companies pay supermarkets marketing allotments or "slotting fees" to ensure optimal shelf space or store location. These fees buy the looming structure of sodas or chips at the valuable end cap—the end of an aisle—or the row of branded snack foods on a shelf at eye level.

It doesn't stop there. Nearly three-quarters of all middle schools and just about every high school in the nation has vending machines, stores, or snack bars, and the most common snacks are soda, imitation fruit juices, candy, chips, cookies, and snack cakes.

The news about school lunches is equally grim. The government spends nearly $1 billion to buy commodities from farmers each year and gives them to the National School Lunch Program, which is supposed to provide healthy, affordable meals to kids. Schools get free food, farmers are ensured regular income, and the government helps prevent prices from sagging.

Doesn't sound that bad, does it? But according to a recent

study by the USDA, *school lunches routinely fail the government's own nutritional standards,* even when the food is being provided by the government. In fact, fewer than one-third of public schools offered lunches that met the USDA standards for total or saturated fat.

And the government's role in supporting Big Food is not limited to school lunches. It's enormous, and mostly disappointing.

Does the Government Help or Hurt?

Overproduction and marketing drive excessive consumption: The food is available; we're bombarded with "information" about why we should be craving it, or why we'll be healthier or sexier or smarter or richer if we eat it, and we respond by purchasing and eating more.

But why *processed* food? Why not *whole* food, real food, basic food? Why not corn instead of corn chips? Yogurt instead of yogurt bars? Chicken instead of McNuggets? Water instead of Coke or Gatorade?

It's easy enough to scorn Big Food or specific individuals who—compelled by greed or simply opportunity—have exploited land, air, water, animals, and humans in order to make their fortune.

But societies evolved to provide mutual protection. And in theory at least, a democratic government exists to protect the citizens who elect it—not just from external threats

The USDA has consistently favored individual and corporate profits over public health.

but from internal ones. This makes looking at our government's role in food policy especially painful. Much of the blame goes to the Department of Agriculture (USDA), which is responsible for administering the farm bill, the common name for a well-established group of laws that set agricultural policy. (The most recent version is "The Food Conservation and Energy Act of 2008.") This sweeping legislation largely determines what food is grown and also—in theory—provides us with nutritional guidance.

There's a serious conflict of interest there, and it costs the public plenty. For something like 70 years the USDA has consistently favored corporate profits over public health, producers over consumers, and—in effect—lies over truth. Public food policy has not been all bad, but by and large the USDA has not done right by the majority of the citizens.

It's no exaggeration to say that public health has been sacrificed at the altar of profits, and that current food policy, although not quite the worst it's ever been, is sorely lacking. (The current farm bill, just passed by Congress as of this writing, is not much better than its predecessors. Perhaps by the next go-round—in 2012 or 2017, depending on what's determined now—we'll have another shot, and by then public opinion may be strong enough to bring about real changes.)

A very short history of American nutrition advice

Starting in the mid-nineteenth century, when the British government mandated that seamen be given lime juice (hence the nickname "limeys") to prevent scurvy, there have been attempts by government to set dietary guidelines. The USDA published its first recommendations in 1894; the author, W. O. Atwater, advised moderation: "The evils of overeating may not be felt at once, but sooner or later they are sure to appear—perhaps in an excessive amount of fatty tissue, perhaps in general debility, perhaps in actual disease."

Smart guy; things went backward from there. By 1916, the USDA had divided foods into five groups: milk and meat, cereal, fruit and vegetables, fats and fatty foods, and sugars and sugary foods. In the 1930s, the USDA was studying the nutritional requirements of Americans, and came out with the precursor of today's recommended dietary allowances (RDAs). The USDA added 50 percent to what it believed was the average requirement for "normal adult maintenance," and thus set the stage for recommending what became a national habit: overeating. By the mid-1950s, we were presented with the classic "basic four" food groups: milk, meat, fruit and vegetables, and grains.

We take this grouping for granted now, but in fact it was arbitrary. (Why, for example, are dairy and meat different, but not meat and fish and not fruit and vegetables?) By giving equal weight to these four groups, the USDA was assuring all farmers that their interests would be promoted; and by making distinct categories for milk and meat, the agency made certain not only that the two biggest lobbies would be satisfied but that Americans would "understand" the "importance" of consuming plenty of each. Evidently no one important was making money from fishing.

The USDA declined to address the single most destructive element in the American diet: sugar.

And by ignoring sugar (did you notice?) the USDA declined to address what has become the single most destructive element in the American diet. By lumping together fruits and vegetables, by making no distinctions among carbohydrates (whole grains, cookies, and white bread were all the same as far as the USDA was concerned), it downplayed the critical differences among these kinds of carbohydrates, and ignored the distinctions among carbohydrates, meat, and dairy products.

Less than 20 years later, the government blew it again. In the 1970s, the Senate Select Committee on Nutrition and

The kind of fat we eat is far more important than the overall amount.

Human Needs convened to eliminate malnutrition and declared that there was also a need to confront excess in the American diet. That mission was met and aggressively opposed, as you might guess, by the agricultural and food industries, whose interest lay not in confronting excess but rather in promoting it.

So in 1977, when the committee released "Dietary Goals for the United States," the battle lines were clear. The committee found—or seemed to have found—that eating too much of certain foods, specifically meat, was linked to chronic illness. It intended to report this, and to recommend an overall reduction in meat eating.

No sooner had the recommendations been written than the lobbyists came out in force, and before the recommendations could even be made public they were revised. So what the committee reported, in what appeared to be a compromise but has since proved to be a victory for Big Food and a defeat for public health, was that Americans should avoid fat in general and saturated fat in particular. Thus the committee created the now familiar mantra: Keep total fat to less than 30 percent of caloric intake; concentrate on polyunsaturated fat; decrease cholesterol intake, and so on.

Thirty years later, it seems safer to say this:

- We need fat to live, and the kind of fat we eat is far more important than the overall amount.
- It's not entirely clear that either saturated fat or cholesterol on its own is harmful.
- When polyunsaturated fat is hydrogenated, as it is in margarine and shortening, it's dangerous to our health.
- Monounsaturated fats, like those in olive oil, are accepted as the best ones to consume for heart health.
- The only foods whose value seems unquestioned, and which can do you no harm (unless they're bathed in

pesticides, of course), are what we think of as plants—vegetables, fruits, legumes, nuts, and grains.

Nutrition advice meets food policy

At the same time as the Committee on Nutrition and Human Needs was working on its report, changes in government farm policy encouraged boosting meat production as well as that of the grains that were increasingly used to feed livestock. Grain could be much more easily raised for profit than grass, even though cattle are natural grazers ("graze" and "grass" have the same root) and their stomachs cannot easily digest grain.

Back in the 1930s, the Roosevelt administration established a program to help people who made their living off the land cope with the falling prices that resulted from overproduction. The Commodity Credit Corporation, as it was called, set a target price based on the cost of production for corn and other storable commodities. When the market fell below the target price, farmers could take out a loan for the value of the crop, using the crop as collateral. They would then store the corn until the market rebounded, and use the proceeds to repay the loan. If the market didn't improve, the government bought the grain and sold it when prices recovered.

Changes in government farm policy have encouraged increased meat production for at least 35 YEARS.

In theory, the program had several benefits. It regulated production to prevent surpluses, and it created a national grain reserve that could stabilize prices during droughts or other disasters. It also kept farmers from abandoning their land as had happened widely during the early 1930s, when a couple of bumper years that lowered prices were followed by a series of droughts in which crops could barely be grown.

Fast-forward to 1973, when President Nixon made a deal to sell grain to the Soviet Union just when a bad growing season was about to produce domestic shortages. The

Agricultural subsidies cost taxpayers $19 billion a year and benefit only 3100 farmers.

result was prices so high that many consumers boycotted meat. Secretary of Agriculture Earl Butz, in an attempt to drive prices down again, encouraged farmers to plant "fence row to fence row" while replacing some loans with direct payments. In other words, farmers were encouraged to grow wheat, feed crops, cotton, and other designated commodities, then guaranteed payment for them.

Thanks to this complicated system of price supports, loans, and deficiency payments (the government even paid producers the "balance" when market prices fell below preset levels) subsidies have evolved beyond their original intent into farm bills that require months of congressional haggling every few years.

This was a recipe not only for overproducing like mad but for making some farmers rich. Today, this program costs taxpayers about $19 billion a year, and benefits only about 3,100 farmers, mostly big growers of—guess what—corn, soy, and other subsidized crops.

The force-feeding of America

Someone had to eat all that food, and it wasn't the Russians. In fact, it wasn't being offered to them, because producers found that the way to make the most money possible was to grow grain that would not be eaten directly (we don't eat much grain anyway) but processed into profitable and easily transportable things like animal feed, white flour, high fructose corn syrup, and oil. (Michael Pollan, in a depressing, amusing, and brilliant article written for *The New York Times Magazine* in 2003, reminds us that surplus grain was once converted to alcohol; now, in a nation that would rather eat than drink, much of it is converted to sugar and oil.)

Thus commodity foods have remained artificially cheap. As a result, we're eating about 25 percent more calories per day than we were in 1970. We're also eating about 10

Consumption Rates of Various Foods (in pounds per capita)

	1950–59	1960–69	1970–79	1980–89	1990–99	2000
Meat, chicken, and seafood	138.2	161.7	177.2	182.2	189	195.2
Fats and oils	44.6	47.8	53.4	60.8	65.5	74.5
Wheat flour, corn, and rice	155.4	142.5	138.2	157.4	190.6	199.9
Total caloric sweeteners	109.6	114.4	123.7	126.5	145.9	152.4

percent more meat, 45 percent more grains (mostly refined), and about 23 percent more sugar.

None of this should surprise anyone. We're simply eating more, and though people may argue among themselves that the type of calories matters, not a single expert will question that excess calories incline you to gain weight.

We have looked at some of the factors behind this phenomenon—oversupply and marketing chief among them—and there are others, all of which surged in the 1970s:

- The decline of home cooking, accompanied by a dramatic upswing in restaurant dining. According to the USDA, during the 1970s 18 percent of Americans' daily calories came from food eaten outside the home. By the mid-1990s this number was 32 percent—nearly double.
- Thanks to USDA recommendations, people who grew up without ever eating a fresh vegetable (unless you count French fries) thought Snackwells were healthy because they were low-fat.
- Meat has become a convenience food (there are barely bones anymore).

By the seventies home cooking was in such a sad state that fast food was more flavorful.

- Precooked food from supermarkets and restaurants has become prevalent.
- With masses of women entering the workforce since the 1970s—and cooking not enough of a priority for men to begin sharing the burden—ease and speed of preparation has taken precedence in choosing what to cook and eat.

Who sees meals with home-cooked breads, desserts, and soups, for example? All these can instead be bought reliably at any store. (Not that they're good, but they're there.) Even salads and precooked main courses are now common take-out foods.

It's worth noting that by the 1970s home cooking was in such a sad state that fast food was considerably more flavorful than what most people were cooking at home. The high fat, salt, and spice content of fast food made it all the more appealing.

The USDA did little to stop these trends. In fact, it was one of the arms of the government that actively participated—whether intentionally or not—in their promotion.

The evolution of the food pyramid

All along, the USDA has walked a tightrope, subsidizing and supporting Big Food on the one hand and trying to fulfill its obligations to American citizens on the other. In 1992, its dietary guidelines (Congress mandated that these be updated every five years, and approved by a panel that includes representatives of the food industry) took the now familiar shape of a pyramid, the implication being that one would eat more from the broad base and less from the narrow top. (It is shown on the next page.)

The pyramid was attacked immediately. Those who maintained that fat intake was too high were outraged that up to six servings of meat and dairy per day were recom-

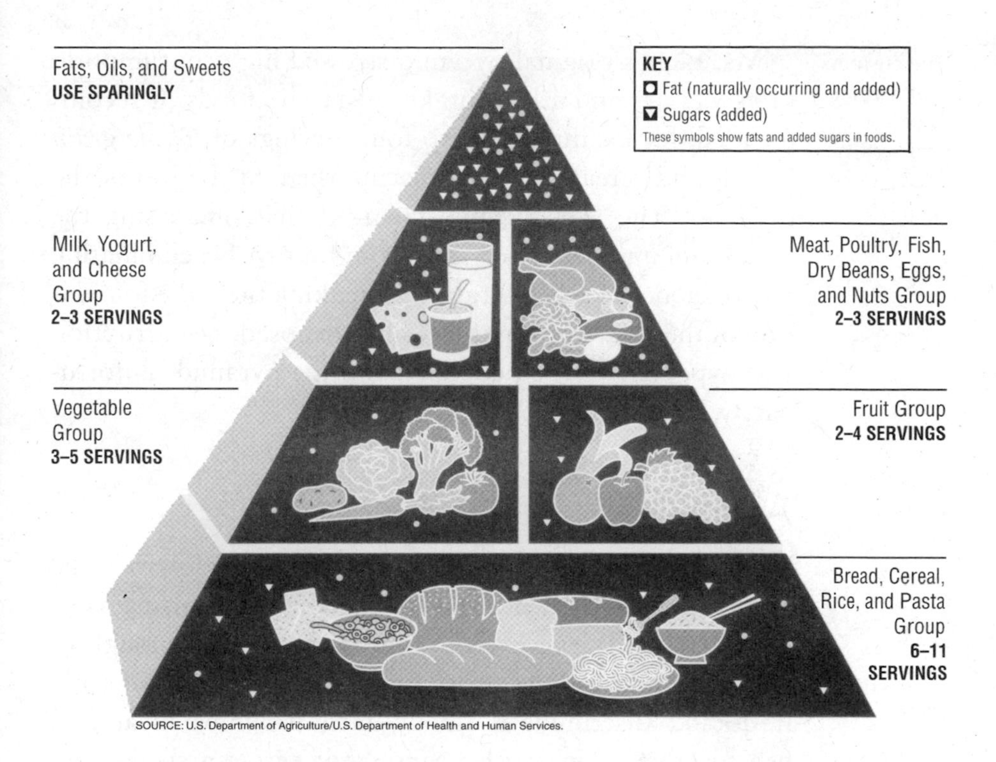

SOURCE: U.S. Department of Agriculture/U.S. Department of Health and Human Services.

mended. Those who thought vegetable and fruit intake was too low were incredulous that as few as five servings per day were considered adequate.

And almost every expert outside the industry that would profit from the recommendation found the carbohydrate grouping insane. Nowhere did the pyramid suggest that any of the servings of bread, cereal, rice, and pasta contain whole grains. Every serving could—at least in theory—be highly refined, so that they'd eventually be metabolized just like the sweets that were to be eaten "sparingly."

As it happened, this was one of those "Our government would never do this, would it?" situations, in which the answer, sadly, is often "Yes." The pyramid had been shanghaied from its original designer, a nutrition expert at New York University, Luise Light, who consulted on its development for the USDA.

The current food pyramid is so vague and confusing as to be nearly worthless.

Ms. Light's original pyramid stressed high consumption of vegetables and a low intake of starchy foods. It recommended, for example, two to four servings of *whole grain* breads and cereals per day, placing them at the top of the pyramid. The USDA panel changed that, increasing the maximum number of servings from four to 11, eliminating any reference to whole grain, and making these the foundation of the diet. Ms. Light has since exposed the corruption and hypocrisy of the people behind the pyramid (unfortunately, though, with not much effect).

A look at today's pyramid

The current food pyramid, released in 2005, is in some ways more sensible; but it is so vague and confusing as to be nearly worthless, even if you are motivated enough to go online and follow all its links (www.mypyramid.gov). It doesn't discriminate among fats; it lumps meat and fish together, along with beans, as if eating a steak were the same thing, health-wise, as eating a piece of flounder or a bowl of chickpeas; it doesn't recommend that any food should be avoided or even minimized; it suggests that half the grains you consume be whole grains, which is the same as saying half should be refined, the kind of carbohydrates that increase the risk of diabetes and heart disease.

And it's about as anti-intuitive as could be imagined. For example, if you seek advice on how much is needed from the "meat and beans" group, you will read:

> The amount of food from the Meat and Beans Group you need to eat depends on age, sex, and level of physical activity. Most Americans eat enough food from this group, but need to make leaner and more varied selections of these foods. Recommended daily amounts are shown in the chart.

SOURCE: U.S. Department of Agriculture/U.S. Department of Health and Human Services.

When I go to the chart I find that a man my age should be eating about 2.5 ounces of meat per day. Not bad; it's probably about a third of what most of us currently eat. You'd think that a government agency recommending that Americans cut their meat consumption by more than half would be raising quite a row; but not this government agency, because the USDA doesn't *really* want this to happen.

The pyramid recommends a higher intake of dairy (including ice cream) than of vegetables.

Now when I look at dairy, I see that a man my age should be eating three cups of dairy "equivalent" a day (including, if I like, one and a half cups of ice cream), more than my recommended amount of two and a half cups of vegetables.

We would have to look quite hard to find an independent nutritionist or nutrition scientist who believes Americans need to eat more dairy than vegetables, especially in the light of recent findings that plants are probably the most beneficial foods you can eat, and that too much dairy at any age, including childhood (with the exception of breast milk), is worse than not enough. Furthermore, *up to 50 million Americans cannot tolerate any dairy in their diet at all.*

The pyramid can be picked apart for many other reasons. There are some non-USDA pyramids that actually make sense—see, for example, the one for the Okinawa Diet on www.okinawaprogram.com—but none is available to anyone without a strong motivation to search. Also, the USDA chose the global public relations company Porter Novelli, which has worked for McDonald's and the Snack Food Association, to design the pyramid itself. In addition, if you're not computer-savvy the pyramid is virtually impossible to find. And most recently, the USDA has announced that food companies can apply for permission to use the logo on their packaging and marketing materials.

But such sniping is futile: the USDA is simply too easy a target. Even assuming that the science behind the pyramid was accurate, and that nutrition guidance by committee can work, none of this would matter—because the committee's thirteen members include seven who have ties to food or drug companies (or both) or have received funding from such companies. (One member has petitioned the FDA to exclude refined sugars from food labels!)

The result is that despite the undoubtedly hard work of its more well-intentioned creators—who have at least encouraged us to limit our intake of sugar, meat, and overall calories and have emphasized the benefits of fruits, vegetables, and whole grains—the food pyramid has evolved into a wildly confusing graphic, generating a flood of criticism.

And this is the major public health vehicle—with a paltry $55 million budget—to counter the billions spent by Big Food in its marketing efforts.

Wait. There's more.

C
H
H
H
H
H

So-Called Healthy Ingredients

As bad as it is, the pyramid is among the most positive things the federal government has done in public nutrition education in the last generation or so. Among the tragedies is the Nutrition Labeling and Education Act of 1990, which, along with mandating nutritional labeling on most packaged food (generally a good thing, though even that is arguable), authorizes the use of (in the FDA's wording) "claims about the relationship between a nutrient or food and a disease or health-related condition," such as calcium and osteoporosis, and fat and cancer.

While the FDA claims that this allows consumers to make informed, intelligent choices, the reality is quite the opposite, a large-scale scam that allows packagers of processed foods to toss, say, a little calcium or soy in with their largely nonnutritive foods and claim that these foods "have the potential to prevent osteoporosis" or "reduce the risk of heart disease."

This also allows manufacturers to use terms like "enriched wheat flour," usually meaning flour from which

The smart consumer buys few foods that have more than one ingredient.

nearly every nutrient has been stripped and to which is added a variety of chemically produced vitamins or minerals, leaving micronutrients in the dust. They can label foods "zero trans fat" even when they're loaded with other fats. Or slap the ubiquitous "all-natural" label on things, though it means almost nothing. The smart consumer pays no attention to any of this. In fact, the smart consumer buys few foods that have more than one single ingredient, and almost no foods that make these kinds of ridiculous dietary claims. (As Michael Pollan says, "A health claim on a food product is a good indication that it's not really food.")

That profit is the motive behind pushing the wrong diet to Americans isn't surprising. That government hasn't protected our best interests, and that the Big Food complex of agribusiness and food manufacturers doesn't have our best interests at heart—perhaps, sadly, this is about what most Americans have come to expect.

"A health claim on a food product is a good indication that it's not really food." —MICHAEL POLLAN

In addition to this confusion, the American consumer is faced with a barrage of conflicting studies about the link between diet and health, studies that Big Food often uses as a marketing tool. And it should come as no shock that Big Food, along with the pharmaceutical industry and its scientists for hire, has promoted confusion in the media and in the mind of the American consumer to contribute to our culture of overconsumption.

Is nutrition science?

It's not entirely malicious. Human nutrition is as complicated as quantum physics, and experiments involving human subjects have some unique challenges compared with those involving electrons: you cannot, after all, ethically deprive people of nutrients or deliberately fill them with empty calories. Unfortunately, this means that we are all guinea pigs in a debate among studies ranging from the well-intentioned to those that blatantly promote one food or food component over another.

It's the nature of science for theories to be disproved. Anyone who even casually follows physics knows that the knowledge of the last decade or the last year is frequently challenged and often even replaced by more recent discoveries.

But though physicists have their jealousies and competitions, and money is often at stake, it's usually small potatoes—not a lot of people read about or make money on muons—compared with what's at stake in "discovering" a "new" nutrient that can be marketed to the American public as The Next Big Thing.

Indeed, nutritionism—a term coined by an Australian sociologist, Gyorgy Scrinis; and popularized by Michael Pollan—is big money and big news. Barely a day goes by without the media calling our attention to a study claiming to have discovered a nutrient that fights cancer, heart disease, Alzheimer's, or another fearsome disease.

Often, within a year that nutrient fades from public view or is discredited. Meanwhile, Big Food has moved into action and "enriched" packaged food—with "whole grains" (most often misleading), or flaxseed, or omega-3, or the latest exotic import. (As of this writing, that would be the acai berry, with more antioxidants—I just read—than any other natural source. Don't rush out to buy it.)

There's nothing wrong with proposing theories and finding results only to have them disproved. The scientific method, as we learned in high school, consists of asking a question, forming a hypothesis—a possible answer to the question—testing the hypothesis, and drawing a conclusion. Sometimes the conclusion is found to be "correct," though even that does not mean it's proven conclusively.

In the 1960s, sufficient evidence was gathered to support the hypothesis that smoking is a cause of lung cancer. There have been no studies (except for a few funded by the tobacco industry) that have ever given any evidence to the contrary. But obviously, not everyone who smokes develops lung cancer (nor is everyone who develops lung cancer a smoker). So there is a correlation, not a perfect cause-and-effect relationship.

Perhaps in part as a result of the enthusiasm those studies generated it was declared that saturated fat caused heart disease. Now there may in fact be a correlation here, but it's nowhere near as strong as that between tobacco and lung cancer. However, the link was being described as cause and effect, a far stronger one.

But this has never been proved; rather, all you can say with assurance is that *something* in the American diet causes obesity and increases the rate of heart disease. It's an epidemic, ultimately affecting more than half the population, and it's not found in anything approaching this frequency in countries where people follow more traditional diets like those of Italy, Japan, China, Mexico, India—you name it—no matter where you look.

All that can be said is that *something* in the American diet causes obesity and increases heart disease.

In any case, saturated fat is just part of the American diet, not the entire American diet. And though saturated fat *may* be a part of the problem, the problem is certainly not entirely the result of too much saturated fat.

In retrospect, it appears that the wise advice, 30 years ago, would simply have been, "Eat fewer animal products and nutrition-poor—i.e., junk—food; eat more plants." This is a simpler message, and far easier to understand, than, "Reduce saturated fats and keep total fat consumption to less than 30 percent of your total calories," which not only required far more understanding of nutrition than most people had, but also required a calculator. *And* it led people to believe that as long as food was low in saturated or total fat, they could eat as much of it as they wanted to, thus measurably eating more calories, and measurably increasing the rates of obesity, diabetes, and heart disease—at a minimum.

By encouraging the consumption of more meat (most people who cut back on saturated fat did so by substituting chicken for beef, and for a variety of reasons—quite possibly because they believe it to be "healthier"—they ate more), of refined carbohydrates, and of trans fats and chemically extracted oils (all low in saturated fats), the government-endorsed dietary model may have led to more deaths than it claims to have prevented. If, during the last 40 years, the government had been promoting a diet of less meat and refined carbohydrates and more vegetables, fruits, legumes, and whole grains, we'd unquestionably be a healthier country.

If the government had promoted a diet of less meat and refined carbohydrates and more vegetables, we'd be a healthier country.

In fact the risk factors for heart disease and the incidence of many cancers continue to rise, and cardiovascular disease remains the number one cause of death. The Central Intelligence Agency estimates the rank of U.S. life expectancy at 46th in the world—behind Jordan, South Korea, and Bosnia and Herzegovina.

Since as a country we do eat less saturated fat than we used to, we can probably say that this kind of fat is *not* the sole culprit. A more likely candidate is the typical American diet.

Yet if it's so simple—"Don't eat the modern American diet, and your health will improve" (in fact it *is* almost that simple)—then why do so many recent nutritional studies do little more than confuse us?

Let's study which diets work

Most studies about diet are contradictory, and definitive results are rare. To illustrate this, let's take a look at the well-publicized Nurses' Health Study and some work done on the Atkins diet, both in the mainstream, both well-intentioned, and both—as far as I can tell—examples of research that in the long run may forward the cause of knowledge but in the short run don't give us much guidance.

Studies have not even demonstrated conclusively that low-fat diets aid in weight loss.

The Women's Health Initiative (WHI) Dietary Modification Trial is part of the Harvard Nurses' Health Study, begun in 1978 and among the longest-lasting of health studies. Starting in 1993, the WHI trial followed nearly 50,000 women in an attempt to demonstrate the health benefits of a low-fat diet.

It did no such thing. Rather, it found that women on low-fat diets did not have lower rates of breast cancer, colorectal cancer (thought to have been at least partly caused by excessive fat intake), or cardiovascular disease.

The study did not even demonstrate that low-fat diets aided in weight loss: women on the WHI diet generally weighed the same as women who followed their usual diets. (Many people believe that women on the diet "cheated," and of course cheating is not uncommon in such situations. But this is not a criticism of anyone: You cannot monitor food intake constantly, or force people to act against their will. Again, this is among the core problems with studies in nutrition.)

All this came as a rude awakening to the health and diet industry, which had expected this large study to confirm its well-publicized beliefs, beliefs previously substantiated by other studies, including the Women's Intervention Nutrition Study (WINS), which made news when the researchers reported that a low-fat diet may help prevent breast cancer. In this study, 2,437 women who been treated for breast cancer maintained either a low-fat diet (their consumption of fat was 20 percent of their daily calories) or a standard diet. The low-fat group showed a 24 percent reduction in the risk of a recurrence of the cancer.

So, which result is "true"? Both are, or neither is. More studies are needed. In the long run—which may mean 20, 50, or 200 years—whether diet can reduce the risk of cancer will become clear. But it isn't clear now.

Let's look at a study of a diet that may be thought of as the anti-carbohydrate and essentially pro-fat diet—most often called the Atkins diet, after the doctor who popularized it. In 2002, in a study at Duke University, some participants were fed 60 percent of their calories as fat while their carbohydrate intake was kept low; the other participants followed the American Heart Association diet—low in fat and relatively high in carbs. Those on the Atkins diet lost more weight and—stunningly—had better cholesterol overall.

But five years later, a study at the University of Maryland Medical Center found that "people on the Atkins diet had increased levels of LDL ('bad') cholesterol," and these researchers declared that the maintenance phase of the Atkins diet not only led to no weight loss but was "potentially detrimental" for cardiovascular health.

Confused? You should be. And it's worse than it appears. There are those who argue that the WHI diet was not even a real low-fat diet, and that therefore its results don't tell us much. And others argue that the markers used to measure the success or failure of *both* diets are irrelevant, and therefore that neither diet tells us much about anything.

Up to 2/3 of people on diets regain more weight than they initially lost.

It may be that extreme diets are themselves unhealthy.

In fact, it may be that extreme diets—at least those used for weight loss—are themselves unhealthy. A recent study at UCLA, the most comprehensive and rigorous analysis of diet studies to date, found that people on diets typically lose 5 to 10 percent of their weight in the first six months, but that within four or five years between one-third and two-thirds of them regain more weight than they lost. Also, there's evidence to suggest that repeatedly losing and gaining weight is *itself* linked to cardiovascular disease, stroke, diabetes, and altered immune function.

Not only do many studies contradict one another; the hopeless consumer is led by the nose, following the diet of the year or looking for the nutrient of the month. After 20 years of being told that polyunsaturated fat was the key to good health, we were then told that the hydrogenated forms, known as trans fat because of the trans fatty acids that form their chemical structure, are in fact damaging. This means that more than 30 years after first hearing that butter was "bad," we learned that the common substitutes like margarines and vegetable shortening were likely to be worse.

After a series of studies "determined" that oat bran lowered cholesterol (remember, it's not even clear that cholesterol is "bad"), oat bran was marketed as if it were the greatest thing since the Salk vaccine. But a more careful study found that high-fiber diets didn't directly reduce cholesterol. Rather, it seemed, that by eating oat bran people simply ate less of foods that could be damaging. Similarly, olive oil itself probably has no benefit other than that people use it in place of more harmful fats.

Olive oil and oat bran, for example, may indeed have beneficial effects, but perhaps these show up in studies because by eating them people are not eating something that might be damaging, and this is important.

It could well be—and this is as close as I can get to Something I'm Very Nearly Sure Of—that by eating simple, natural, minimally processed foods, known to be at least benign if not beneficial, in place of foods that are suspect in any quantity (junk food, highly processed carbohydrates), or those that may be damaging if consumed in large quantities (animal products), you're going to be healthier and quite likely thinner. And if you believe me, you don't need to follow the results of any more studies.

So why fund studies?

What good, then, are nutritional studies? Undoubtedly they move knowledge forward, however slowly and falteringly. Were it not for these studies, journalists, researchers, and nutritionists would not be able to see emerging patterns. (Each person may draw different conclusions, but the best studies and the most knowledgeable commentators are currently all pointing in a similar direction.)

But since many studies are funded by the same people who raise, manufacture, and market the products being researched, they must often be viewed skeptically. The potential benefits to sponsors in discovering that their food is "healthy" are great.

Studies of soft drinks, juice, and milk had a better likelihood of favorable outcomes when funded by the food industry.

The commodity boards discussed in Selling the Bounty (page 31) usually require their members to contribute to research as well as promotion. So we see studies like these:

- The United Soybean Board funded research that established a link between consumption of soy protein and a reduced risk of heart disease. (In 1999, this discovery led to the Food and Drug Administration's establishment of a health claim regarding the cardiovascular benefits of soy protein.) Now, the same board is partially funding a study by the National Institutes of Health (NIH) into the role of soy in preventing prostate cancer.
- The California Walnut Board helped pay for a study showing that walnuts may be better for heart health than olive oil.
- The Almond Board of California and Unilever (manufacturer of Take Control margarine) were behind a study to show the cholesterol-lowering benefits of foods like raw almonds, tofu, and margarine enriched with plant sterol—like Take Control.

Like many studies, these found what they were looking for. That's not surprising: A study showed that research on soft drinks, juice, and milk had a better likelihood of favorable outcomes when funded by the food industry. And if studies don't find exactly what their sponsors want, there are ways of dealing with that, too:

- In 2007, a look at studies funded by a single drug company found that they had a 55 percent rate of getting the results desired by their funders, but that they were reported positively 92 percent of the time. (This gap vanishes entirely when studies are performed by nonprofit institutions.)
- A study at the University of North Carolina revealed that studies of breast cancer therapies are more likely to report positive results when funded by the pharmaceutical industry than when funded by other sources. This review showed that industry-funded studies are *designed* for positive outcomes, often looking only at how a drug performed, without a control group for comparison. (Many studies with control groups taking placebos—substances that have no medical effects, like sugar pills—find the placebo as effective as the drug being tested.)
- Researchers at the University of California–San Francisco found that when trials compared two cholesterol-lowering drugs, the results were *20 times* more likely to favor the drug from the company that funded the study.
- A story in the *New England Journal of Medicine* reports that 94 percent of positive studies funded by drug companies find their way into print. The number for those with ambiguous or negative results?—14 percent.

Olive oil, walnuts, and acai berry have taken advantage of their status as magic bullets to raise sales.

The disconnect between data and reporting is called spin. And people believe it.

Olive oil, walnuts, and acai berry—all intrinsically healthful foods—are good examples of foods that have taken advantage of their status as magic bullets to raise sales. Olive oil consumption has nearly tripled since 1991; walnut consumption reached a record high in 2004/05, of 0.54 pounds per person; and the acai berry has seen sales triple in just three years, from $3.8 million in 2005 to $13.5 million in 2007.

The health industry continues to promote consumption and "cure" over lifestyle modification.

Why? Because we've been led to believe that there may be a magic bullet for our health problems. In general the health industry continues to promote consumption and "cure" over straightforward lifestyle modification. Whether it's taking a pill to lower cholesterol or eating acai berries, the promise of an easy solution trumps eating and living simply and sensibly.

The easiest, surest way to improve the overall health of Americans is for us to adjust our eating habits.

The food and pharmaceutical industries benefit mightily from this belief, so it throws literally billions of marketing dollars a year at reinforcing the behavior. (In 2007, all the potential "blockbuster" drugs—those with potential sales of $1 billion or more per year—were designed to combat lifestyle diseases: diabetes, heart disease, and obesity.)

The value of these studies would diminish substantially if we ignored them. But with the print, broadcast, cable, and Web media competing for our attention, every bit of "news" is treated as earth-shattering, even if the sources are questionable. There is no news value in saying, "Eat a sensible diet," whereas there is enormous news value in saying, "New omega-3 research shows cancer reduction"—followed, quite likely, by a commercial for soy milk fortified with omega-3s.

There is no magic bullet

The easiest, surest way to improve the overall health of Americans is for us to adjust our eating habits. We're not talking about a diet. It's a change in focus, away from the twentieth-century style of eating and back to something saner, more traditional, and less manufactured.

It's indisputable that our excessive food intake is bad for our health. The full effect of obesity on overall health has not been determined, but at the very least it's responsible

Almost

1/4

of Americans have some form of cardiovascular disease, and nearly as many have diabetes or pre-diabetes.

for reduced mobility and circulation, and for stress on several organs. And there's no question that we're getting fatter, so much so that there are people who use the term "obesity epidemic." There are twice as many overweight adults as there were 25 years ago, and about three times as many children and teenagers.

Meanwhile, heart disease, cancer, and stroke remain the first, second, and third leading causes of death in the United States. About 70 million Americans (almost one-fourth of the population) have some form of cardiovascular disease, which is responsible for more than 6 million hospitalizations each year. Nearly 21 million Americans suffer from our sixth-biggest killer, diabetes (and six million of them don't know it); and another 41 million have pre-diabetes. It seems absurd to put a cost on these diseases, but people do, and the latest cost estimate for diet-related illnesses—heart disease, stroke, diabetes, obesity, and cancer—is roughly $840 billion, according to the Centers for Disease Control (CDC). (The budget for Social Security, by comparison, is $657 billion.)

Almost everyone believes that these are diet-related diseases. The CDC believes that much of the burden of heart disease and stroke could be eliminated by reducing their major risk factors: high blood pressure, high cholesterol, smoking, diabetes, physical inactivity, and poor nutrition. Others estimate that improved nutrition and lifestyle could reduce illness and death from cancer by as much as 40 percent, death from cardiovascular disease by up to 30 percent, and cases of diabetes by *at least* 50 percent. Those are big numbers, and "improved nutrition and lifestyle" is the approach that *Food Matters* shares.

We have not been moving in the direction of "improved nutrition," though, and consequently we've seen the situation get worse. Since 1990, those diagnosed with diabetes have increased 61 percent; since 1991, the prevalence of obesity has increased 75 percent; and heart disease is not only the number one killer of adults: frighteningly, it's also the second leading cause of death for children under 15.

But if we know what to do, and we know how to do it, why aren't we doing it? The answers are complicated and not entirely clear. For one thing, we're not agreed on what to do. For another, there is a disincentive on the part of the food industry for us to change the way we eat. And for a third, the government agencies that might encourage the "right" diet are beset by confusion and by their ties to the food industry.

It would be unprofitable—for Big Food at least—if we moved our eating habits in the right direction. But not only would profits fall if we ate a moderate diet of wholesome foods; so, too, would the rates of lifestyle diseases, global warming, our collective weight, antibiotic use, environmental damage, and cruelty to livestock. What's stopping this move, largely, is inertia, habit, a lack of good information, and a drive to maintain the status quo by the people who profit from it. But maintaining the status quo is insane.

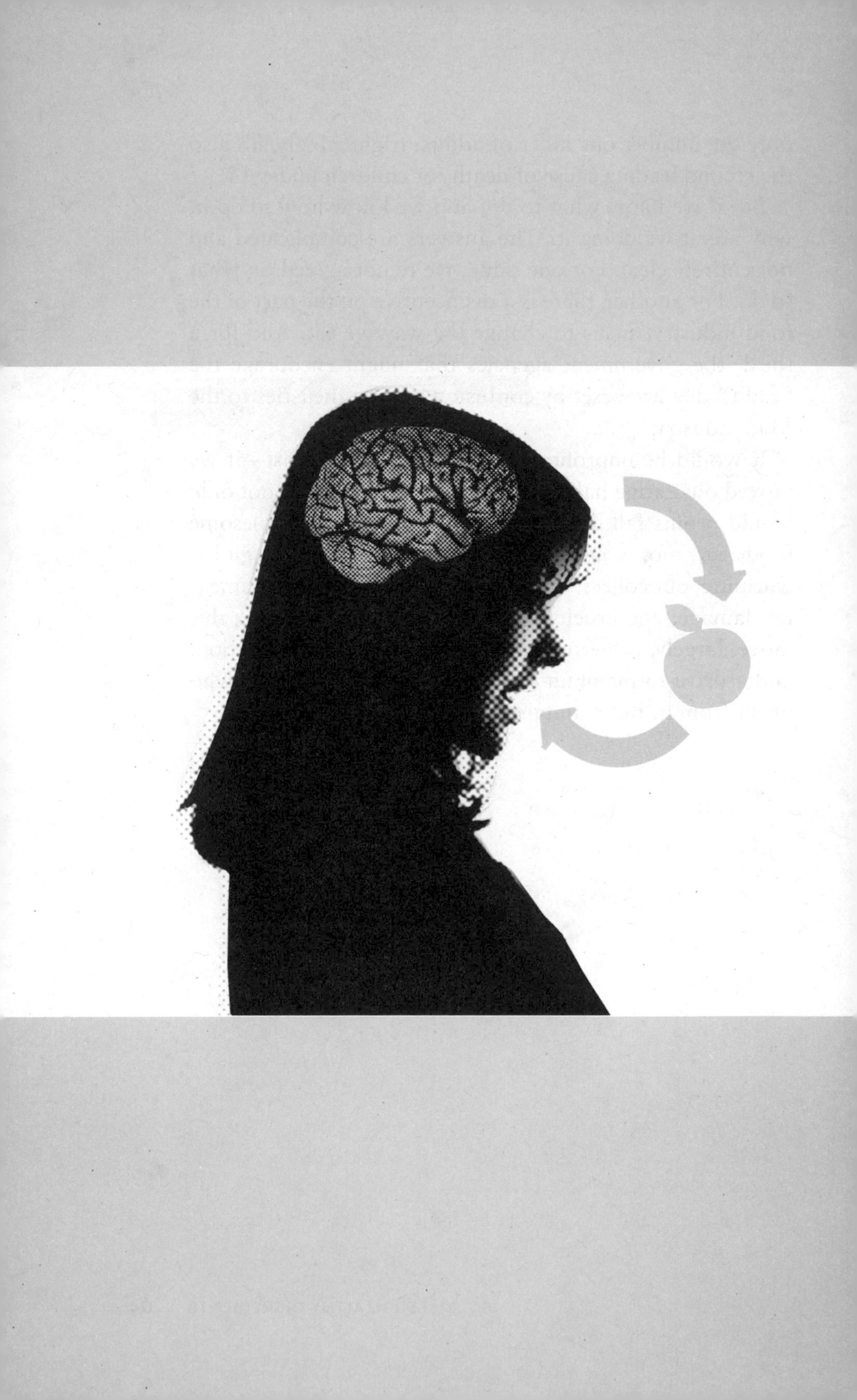

Sane Eating

In sum: Much about the typical American diet is wrong. It's damaging both individually and globally, and we can't expect Big Food or the government to help us fix it.

But the realization of just how straightforwardly and even easily we can make things right—at least a great deal for ourselves, and to some extent for one another—was the driving force behind my decision to change the way I ate. The more I understood about the relationship between human and environmental health, the more I felt a need to act. (As I said in the Introduction, a key moment for me was the publication of *Livestock's Long Shadow,* the UN report revealing the link between raising animals and climate change.)

Equally important, though, since I was unwilling to give up one of life's basic pleasures, was that I saw a way to introduce a much better diet into my own life without much sacrifice.

At first, I simply eliminated as much junk food and over-refined carbs as I could, along with a sizable percentage of animal products. All this turned out to be easy enough, for a couple of reasons. One, when I did allow myself to eat meat, or dairy, eggs, sugar, or bread made from white flour (usually at dinner), I ate whatever I wanted, and as much of it as I wanted. And two, I started to lose weight, quite quickly—a big boost of positive reinforcement.

I wondered: If the cumulative effect of the American diet could have such a negative impact on our bodies and the planet, then couldn't individuals help reverse the damage—again cumulatively—by making small changes in what they choose to eat?

Clearly, the diet was helping *me*; I lost weight and saw my cholesterol and blood sugar improve dramatically. But my impact on the industrial meat and junk-food complex—what I've been calling Big Food—and on slowing climate change was obviously insignificant. Suppose, though, I could get others on this bandwagon? This way of eating is far from complicated, has few rules, makes sense, and works. It can have its own reward in better health and often weight loss, but it also is a way to save energy in the same way as carpooling, turning off the lights when you're not in the room, lowering the thermostat during the winter and wearing a sweater in the house, installing a windmill, whatever other parallel you care to draw.

This way of eating is far from complicated, has few rules, makes sense, and works.

So. Welcome to *Food Matters*: a not very new (but for most Americans novel) way of eating that's personally healthy and globally sane but not deprivation-based, faddist, or elitist. No calorie counting, and no strictly forbidden foods: Just a few quite specific recommendations that you can adapt to your own style.

Sane eating, simplified

Here's the summary: Eat less meat, and fewer animal products in general (I'll get to specifics on page 93). Eat fewer refined carbohydrates, like white bread, cookies, white rice, and pretzels. Eat way less junk food: soda, chips, snack food, candy, and so on. And eat far more vegetables, legumes, fruits, and whole grains—as much as you can.

If you followed those general rules and read no farther, you'd be doing yourself and the earth a favor. And I'm by no means the only one who thinks so.

Shortly after I started eating this way, an article appeared in *Lancet*, the highly respected British medical journal, that supports the general position of *Food Matters*, even in its specifics: "Particular policy attention should be paid to the health risks posed by the rapid worldwide growth in meat consumption, both by exacerbating climate change and by directly contributing to certain diseases."

As a measure of progress, the authors propose this: "The current global average meat consumption is 100 g per person per day, with about a ten-fold variation between high-consuming and low-consuming populations. 90 g per day is proposed as a working global target, shared more evenly, with not more than 50 g per day coming from red meat."

Ninety grams a day is about 3 ounces (50 grams is not even 2 ounces; it's less than an eighth of a pound); Americans' per capita consumption, as I've noted earlier, is more than 8 ounces per day. You might eat more than that; you might eat less. But for most Americans, cutting down to the international average would be a huge step (cutting 10 percent beyond that would be practically insignificant). In fact, it we ate the world average, 3 ounces a day, that average would fall to about 90 grams a day, or just about what *Lancet* recommends.

The goal of eating sanely is not to cut calories; that will happen naturally.

The goal of eating sanely is not to cut calories; that will happen naturally, and you probably won't notice it. The goal is not to cut fat, either; in fact it is possible that you eat more fat than you do now, although different fat. The same is true of carbohydrates—again, you may wind up eating more, but different kinds. And the goal is not to save money, though you will.

No—the goal is simply to eat less of certain foods, specifically animal products, refined carbs, and junk food; and more of others, specifically plants, in close to their natural state.

If you made those your goals, you'd change your life. You'd probably weigh less, you'd have lowered your chance of heart disease and other lifestyle diseases, and you'd make a contribution to slowing global warming.

For a variety of reasons—it's not temporary, no foods are strictly forbidden, and there's no calorie counting—this is not what's popularly called "a diet," as in "I'm on a diet." Rather, it's a shift in perspective or style, an approach.

In any case, the principles are simple: deny nothing; enjoy everything, but eat plants first and most. There's no gimmick, no dogma, no guilt, and no food police.

I want to stress, too, that this is not a new way to eat, but one that's quite old-fashioned; you could even say it's ancient. Among our ancestors, there were few people who did not struggle to get enough calories; it was only in the late twentieth century that people could and did begin to overeat regularly. Until then, most people considered themselves lucky to eat one good meal every day; many people spent half the year eating poorly, and the other half eating decently, though certainly not lavishly, except on certain feast days and holidays. Think of Lent and Mardi Gras, meatless Fridays and Sunday dinners, festivals in autumn and spring, and more. These were all formalized acknowledgments that food was and is something to be celebrated and enjoyed, but overdone only occasionally. *Food Matters* is no more than a way to look at this from a contemporary perspective.

There's no gimmick, no dogma, no guilt, and no food police.

You could say that this way of eating is ancient.

You can go from here to there a number of different ways. You can opt out of two servings of meat a week, or of all but two servings of meat a week. You can eat an apple (or three) instead of potato chips this afternoon. You can start the day with oatmeal instead of bacon and eggs, and so on. You'll find many substitutions, ideas for specific eating styles, and recipes, starting on page 111.

How I got here

My route to saner eating was more or less accidental. Two years ago, I was 57 and weighed more than I ever had before. When I graduated from college, I weighed 165 pounds. When I stopped smoking, about five years later, I weighed 180. When my first daughter was born (and when I started writing about food), I hit 190. Over the next 20 years or so, I managed to gain another 25 pounds or so, until I reached 214. I'm not a small person, so I didn't look that heavy (or that's what I liked to think, though people now tell me otherwise), but you could tell I was overweight, and I developed a number of the expected health problems. My cholesterol was up, as was my blood sugar (and there's diabetes, as well as serious obesity, in my family); I had a hernia; my knees were giving out (your knees know how much you weigh!); and I had developed sleep apnea.

I was also working on *How to Cook Everything Vegetarian*. I had (and still have) no intention of becoming a vegetarian, but I could see the writing on the wall: Industrial meat production had gone beyond distasteful and alienating right through to disgusting and dangerous; traditional, natural ingredients were becoming more and more rare; and respectable scientific studies were all pointing in the same general direction.

For me, the combination of cholesterol, blood sugar, and apnea was the real trigger. I'm not going to go into specific details, but my problems were scary enough and, according to my doctor, all easily remedied. For the cholesterol, I could take statins—cholesterol-lowering drugs—or I could try eating less red meat; for the blood sugar, I could eat fewer sweets; for the apnea, I was told to lose 15 percent of my body weight.

No results were guaranteed. Eating less meat doesn't always lower cholesterol; losing weight doesn't always remedy apnea. But to me the combination of these recom-

A combination of many factors points to a style of eating that simply makes more sense.

mendations, along with the upward trend of my weight, what I'd learned about food over the years, my increasing disgust with the way most meat is grown in this country, the UN report, and more, pointed the way to a style of eating that simply made more sense.

Along with my friend and colleague Kerri Conan, I started eating a diet that was nearly "vegan until six," and at first may sound strict. Until dinnertime, I ate almost no animal products at all (I allowed myself half-and-half or milk in my coffee), no simple carbohydrates (though my coffee often takes sugar), and no junk food. At dinner, at least when I began eating this way, I ate as I always had, sometimes a sizable meal including animal products, bread, dessert, wine, you name it, and sometimes a salad and a bowl of soup—whatever I wanted.

That's just one way to approach this style of eating. And though few unbiased nutritional experts would disapprove of it, it might sound counterintuitive to you. Indeed, the opposite schedule—eating the heaviest meal of the day for lunch or even breakfast—may make more sense with regard to strictly body function. But eating this way suits my particular lifestyle. I detest overly prescriptive diets that are ultimately impossible to follow, and the point of this one, again, is to eat more vegetables, fruits, legumes, and whole grains and less meat, sugar, junk food, and overrefined carbohydrates—and to do so without suffering or giving up all the foods you love. How you go about that probably doesn't matter much.

What does matter is results, and mine were striking. First and probably foremost, I had little trouble eating this way. (It was toughest when I was away from home, but I've figured that out, too; see page 106.) Second, I started feeling and sleeping better. Third, I didn't think much about it for a month or two. It was just one of those things that made sense, like—I don't know—realizing at age 40 that I'd never liked wearing wool, or that excess drinking usually isn't worth the hangover.

A month later, I weighed myself; I'd lost 15 pounds. A month after that, I went to the lab for blood work: both my cholesterol and my blood sugar were down, well into the normal range (my cholesterol had gone from 240 to 180 and Kerri's dropped about 23 percent). My apnea was gone; in fact, for the first time in probably 30 years, I was sleeping through the night and not even snoring.

Within four months, I'd lost more than 35 pounds (Kerri has lost 25 over the course of a year), and was below 180, less than I'd weighed in 30 years. (In fact, of all my diet-related ailments, only my knees didn't respond. Oh, well. One does age, after all.) On a five-week stint in Spain I gained back five pounds; I quickly lost those, but by then I'd stopped counting. My weight has clearly stabilized at a new lower level and—probably more important—I'm at home with this way of eating.

What works for me

Suggestions for specifics about how you might go about eating this way begin on page 119, but this is what I do. I eat about one-third as much meat, dairy, and even fish, as I did a couple of years ago. I eat very little in the way of refined carbohydrates. (However, when there's good white bread on the table at dinner I attack it, and I still eat pasta a couple of times a week.) I eat almost no junk food—by which I mean fast food, candy bars, snack food, and the like—though I allow myself the classic combo of cheeseburger, fries, and Coke every couple of months. I eat probably three or four times as many plants as I ever did, and my guess is that 70 percent or so of my calories come from non-animal sources.

70% or so of my calories come from sources other than animals.

For some people, a shift of 10 percent of calories from

animal to plant may feel significant, though I doubt that; it would be the equivalent of maybe not having chicken on your Caesar salad at lunch but keeping the rest of your diet the same. A person making that kind of shift, along with cutting way back on junk food and refined carbohydrates, might still see positive health changes.

A shift of 50 percent—replacing half your animal calories with plant calories—would be significant for anyone, and would take a conscious ongoing effort. It's not very difficult, but it won't happen automatically. (I'm not suggesting for a second that anyone start counting calories in this manner; but you'll know when you've replaced a significant amount of your animal foods with plants. Everything will feel different.)

If, as Joel Fuhrman suggests in his book *Eat to Live*, you choose to get 90 percent of your calories from plants, you'll be conscious of your diet all the time, and you'll work hard at it. I think this is an extreme alternative, but this kind of diet has actually reversed heart disease in several studies and will probably leave you feeling healthier than you ever imagined, and looking better as well.

What *do* I eat? For breakfast, it's either cereal, like oatmeal or cracked wheat, or fruit salad or sometimes vegetables left over from the previous night. If I have a midmorning snack, it's fruit or nuts. For lunch I'm strict about the type of food I can eat, but I make an effort to eat a lot, trying to get really full on beans, grilled or roasted vegetables, salads, fruit, and maybe some grains; if I'm desperate, I'll eat a little pasta. (If I'm home, it's one or two of those things; if I'm out, I just take everything that looks good at the salad bar, with lemon juice or vinegar and olive oil.)

Mind you, I'm not fiercely strict: I don't pay much attention to whether there's a little cheese on the roasted tomatoes, or whether the cucumber salad is made with sour cream or yogurt. I do steer clear of the bacon bits, the mayo-laden salads, and—needless to say—the grill, the sandwich bar (though as you'll see it's easy to make a delicious all-vegetable sandwich), the taco station, and so on. And I don't drink alcohol at lunch.

I don't want to downplay how much of a change this has been, but it's been nearly painless.

Dinner varies wildly. If I'm eating at home, it might be a salad, bread, and maybe cheese; or soup and bread; or stew, with meat; or a piece of cooked fish with a couple of simple sides. If I go out, there are few limits; I order what I feel like eating, and drink what I feel like drinking. Just in case you're wondering, this might mean a typical steakhouse dinner, a grand dinner at a top restaurant, a few typical dishes at a good Italian place, enough sushi to fill me up—whatever.

But as months of this style of eating turned into years, I found myself front-loading even the grand meals with vegetables, and becoming less interested in the heavier meat dishes that followed. This is an important point: My food choices have changed, even when I go out, and they reflect my mood more than what was surely a habit of focusing on meat, with simple carbs in second place. That balance has shifted.

I don't want to downplay how much of a change this has been for me, but at the same time I want to stress that it's been nearly painless. It's as sensible a routine as I can come up with, but it's personal: I find it easier to make strict changes than to make moderate ones, so I'd rather be supervigilant all day long, then relax at night. (It was also easier for me to stop drinking for a year than it is for me to have fewer than three glasses of wine at dinner. What can I tell you? That's me.)

And again, let me stress that these are my personal rules. They work for me, and if they sound good to you, try them. Whatever plan you wind up with (you'll find several beginning on page 122), the direction is the same: more plants, fewer animals, and as little highly processed food as possible.

Saner eating, better health, and weight loss

This eating pattern has several obvious benefits. By reducing the amount of meat we eat, we can grow and kill fewer animals. That means less environmental damage, including climate change; fewer antibiotics in the water and food supplies; fewer pesticides and herbicides; reduced cruelty; and so on. It also means better health for you.

Reducing the amount of simple carbohydrates (including junk food) has similar ecological effects—less in some ways, more in others. Junk food uses tremendous amounts of packaging, for example. It's likely that cutting back on this kind of eating also improves health, since there's compelling evidence that our high consumption of refined carbohydrates (especially all that high fructose corn syrup) is largely responsible for the marked increase in type 2 diabetes.

The more plants you eat, the less you eat of potentially damaging foods.

As time goes on, we may well discover that increasing the amount of plants you eat is the most important part of this plan. Of course, the more plants you eat, the less you eat of other, potentially damaging foods. In a way, it's addition by subtraction.

And it may be more than that. The micronutrients in plants remain little understood, and their benefits are far from being fully described. For example, it seems quite likely that eating an orange gives you a whole set of nutrients that come along with vitamin C but are far more complex than vitamin C, and eating a carrot provides many more benefits than a dose of beta-carotene. (There's little indication that isolating nutrients, even micronutrients, and taking them as supplements, is a key to good health.)

As I've said, this style of eating can also promote weight loss, and that's of primary importance to many people. The explanation is neither technical nor complicated, and mostly centers on the concept of caloric density, made popular by books like *The Pritikin Principle* and *The Okinawa Diet*. The idea is to rely on foods that have relatively few calories by *volume*.

The idea is to rely on foods that have relatively few calories by *volume*.

Think of it this way: From the dawn of human life until the twentieth century, most people had to struggle to get enough calories, so calorie-dense foods were the most highly prized. These included meat, dairy foods, and fats, which, in a well-proportioned diet, are largely beneficial, because they're also among the most nutrient-dense foods. Highly refined grains, sugar, and alcohol (beer, vodka, whiskey, and so on have played important roles in supplying calories in specific cultures) are also calorie-dense, but they're nearly worthless nutritionally, and they are potentially harmful when consumed in large quantities.

If you're struggling to get enough calories, and you want to take in as many calories as you can possibly consume, calorie-dense foods gain in importance. They're also convenient. Though we all love to eat, it takes longer—and takes more work—to fill up on a huge pile of romaine lettuce than on a small steak.

This is in part why I would never argue for a diet that totally eliminates anything. For one thing, such a diet arouses our rebellious streak. For another, it's no guarantee of health; there are plenty of non–meat eaters who get their fill of junk food.

But most important, I think, is that keeping some calorie-dense food in your diet—whether it's meat, pasta, beer, or cake—allows you to reach satiety more quickly and easily. And this will keep you from feeling deprived.

Still, calorie-light foods are the key to sane eating. Most of them—like leafy greens and most other vegetables, brothy soups and stews, fruit, legumes, and whole grains—are full of valuable nutrients and fiber. If you want to be technical, you can calculate the caloric density of any food, by dividing the calories in a portion by its weight (lower is better). For some specifics and examples, see the chart on page 84.

The psychology of weight loss

Of course, willpower is involved in any change of lifestyle. So assuming some will read *Food Matters* with a primary goal of losing weight, it's worth spending a minute on the topic of hunger, weight loss, and will.

There's a basic truth here: there are stages of hunger, and we—Americans in general—have become accustomed to feeding ourselves at the first sign. This is the equivalent of taking a nap every time you get tired, which hardly anyone does.

Eating every time you feel hunger is the equivalent of sleeping every time you feel tired.

There are levels of hunger, and there is a very real difference between hunger and starvation. Starvation is a physical state; your body is deprived of essential nutrients or calories for a long period of time. Probably no one reading this book has ever been truly starving—though we all think we know what starving feels like.

Hunger is a hardwired early-warning system. At first, your brain says, "Think about eating something soon." In the later stages, it says, "Eat as soon as you can; make eating a priority." At no point does your brain say, "Eat now or you will do permanent damage," though at times it may feel as if that is true. But "Eat when hungry" has become a habit. We get hungry. We eat. We get hungry again. We eat again. And so on.

I'm not saying, "Don't eat when you're hungry." I'm saying that if losing or maintaining weight is important to you, think twice before you eat from simple hunger, or from other reasons, like emotion. And when you do eat, choose a piece of fruit; a carrot; a handful of nuts. If you're still hungry, have more. And more. Eat a pint of blueberries, or cherry tomatoes; have a mango, a banana, *and* an apple.

Have a lightly dressed salad. You would be hard-pressed to gain weight eating this way.

Embrace moderation and you will lose weight and tread more lightly on the planet.

You can also embrace hunger, strange as that may sound, just as you might embrace the delicious anticipation of a nap, or sexual craving. Your hunger will, after all, be satisfied; why not wait an hour? (You're not dying, after all!) You might also stop eating before you're full (three-quarters full is probably about right). And if you eat slowly, taking your time, you'll give the food time to reach your stomach and give you a sense of satisfaction before you have seconds or thirds.

If you embrace moderation, eat whole foods instead of junk, live within your physical, monetary, and environmental budget rather than constantly exceeding it, as so many of us do, you will lose weight, tread more lightly on the planet, and gain satisfaction from these things. The next chapter outlines specifics about how you can go about doing so.

How to Eat Like Food Matters

If you're confused by diets and by studies of diets, you may be skeptical about *Food Matters*. My suggestion is that you try sane eating, see how easy it is, and decide whether it works for you.

I assure you that the logic behind *Food Matters* is solid, but you may be curious about why a diet high in plants is so much more desirable than one based on animal products and highly processed foods. (For some readers, that sentence alone will be transparent enough to make the answer obvious.) Here's a brief analysis of the large-scale nutrients in our diet, how they're measured, and the effects they have on your body. If you want to get right down to the business of planning how to eat, turn to page 111.

Defining calories and caloric density

A calorie is the energy required to raise the temperature of one gram of water one degree Celsius. (Usually, when we say one "calorie," we mean 1,000 calories, or a kilocalorie—the amount of energy required to raise a kilogram—1,000 thousand grams—one degree. A food containing 100 "calories" actually contains 100,000 calories, or 100 kilocalories. But we can ignore all this, since it's all relative anyway.) In the process of metabolism, the energy contained in food, calories, is released during digestion, and is in turn used to fuel our bodies.

Excess energy—too many unused calories—may be stored as fat, which can be converted to energy in times of need. (Part of the problem with the typical American diet is that there *are* no times of need for most of us. We are not nomads, polar explorers, or subsistence farmers; rather, we cope with the twin beasts of overproduction and overconsumption.)

100 grams of chocolate cake contains 333 more calories than the same weight of broccoli.

It's common to compare the amounts of calories in food by measuring calories per given weight. Broiled rib eye steak, for example, has about 205 calories per 100 grams; broccoli about 34, and chocolate cake about 367. So if you eat 100 grams of chocolate cake you're eating 333 more calories than if you eat 100 grams of broccoli.

But all calories contain the same energy; as a pound is a pound, a calorie is a calorie. Depending on your size, activity level, metabolism, and so on, you need a certain number of calories to function and to maintain your weight. In theory at least, if you eat more than you need, regardless of the source of the calories—fat, carbohydrate, or protein—you gain weight; if you eat less, you lose weight.

For most of us, the idea is to get the number of calories it takes to maintain weight (or fewer, if we're trying to lose), along with a good balance of nutrients. And this is easy: *As long as your diet isn't based on junk food, almost any diet that supplies you with enough calories will also supply you with adequate nutrition.*

So the idea is to eat food that fills you up (and provides you with nutrients) without giving you more calories than you need. One way to make sure of that is to eat food with low caloric density, and this is less complicated than it sounds—believe me.

This concept, popularized by the authors of *The Okinawa Diet Plan,* is based on the idea that to feel satisfied, most of us need between two and three pounds of food daily. But two pounds of chocolate cake contains 3,330 calories; two pounds of broccoli contains only 309 calories. And we could probably find even more calorie-dense chocolate cake!

Since consuming less than this amount may leave you feeling hungry, the most effective way to lose weight is to rely heavily on foods that have fewer calories per weight; in other words, choose foods with a low caloric density. Simply put, a pound of cake contains more calories than a pound of broccoli. Most foods lie between the two, but you get the idea: calorie-wise, you're better off eating 2 pounds of plants than 2 pounds of junk food, animal food, or refined carbohydrates.

No matter what your diet, as long as it isn't based on junk food, you'll receive adequate nutrition.

There's no need to count calories, but until you get the hang of caloric density, you might want to keep tabs. To do the math yourself, divide the calories in a food by its weight. The lower the number, the lower the caloric density. So broccoli has a caloric density of 0.3; steak, a little more than 2.0; and chocolate cake, 3.7. (For a comprehensive source for all nutritional data, see the USDA database; go to www.usda.gov and search for "Nutrient Data Laboratory." You can look up the values for 100 grams of any food; find the number in the kcal column, move the decimal point over two clicks to the left, and you have the caloric density.)

There's no need to count calories.

Here's a table with some major food types and their caloric density. Take a look, and you'll quickly realize that you can probably estimate almost anything else. Remember—the lower the number, the more you can eat of a food without piling on the calories. It's really common sense: eat

Caloric density of some common foods

Food	Caloric Density
Water, tea, coffee	0.0
Cucumber, lettuce	0.1
Tomatoes, celery, radishes, chard, spinach, summer squash	0.2
Grapefruit, strawberries, button mushrooms, broccoli, bell pepper	0.3
Broth and vegetable soups	0.3
Nonfat milk, carrots, cantaloupe, papaya, peach, winter squash	0.4
Sea greens, hearty greens, oranges and orange juice	0.5
Apples, blueberries, fat-free cottage cheese	0.6
Tofu; tuna canned in water	0.7
Sweet potatoes, potatoes, pasta, most seafood, boneless turkey breast	0.9 to 1.4
Chicken breast, lean red meat, fatty fish, hummus or beans, brown rice and other whole grains, whole wheat bagel	1.7 to 2.0
Vanilla ice cream, skim-milk mozzarella, soy cheeses, low-fat bran muffin, broiled rib eye steak, McDonald's Chicken McNuggets or Filet of Fish, Burger King Double Whopper with Cheese	2.0 to 3.0
Cheesecake, fat-free whole wheat crackers, Swiss or cheddar cheese, air-popped popcorn, glazed doughnut, oatmeal cookie, McDonald's French fries	3.2 to 4.0
Cashews, pistachios	5.7
Bacon	5.8
Peanuts and peanut butter	5.9
Almonds	7.1
Pecans, macadamia nuts, butter, and mayonnaise	7.2
Vegetable oils	8.8

Sources: The Okinawa Diet Plan; *the USDA Nutrient Data Laboratory; McDonald's nutrition facts.*

moderate quantities of foods with moderate densities, and eat small amounts of foods at the high end.

This is only part of sane eating; it doesn't cover the whole picture. But once you add the other principles outlined in *Food Matters*, it's easy to make food choices: emphasize plant foods, and minimize animal products and junk foods with little or no nutritional value, even though their caloric density falls into the same range as that of more healthful whole foods. And even though I advocate plenty of foods with high caloric density, like olive oil and nuts, they're mostly not foods you'd be eating by the cupful, and they're in the larger category of nonmeat, nonjunk, non–refined carbohydrate.

Protein

There are people who will argue that the diet I'm recommending doesn't provide enough protein, or enough complete protein. In part, that's because the meat industry has tried so hard to make "protein" synonymous with "meat," which it most decidedly is not. (Per calorie, cooked spinach has more than twice as much protein as a cheeseburger; lentils have a third more protein than meat loaf with gravy.)

Spinach has more than twice as much protein per calorie as a cheeseburger.

Of course we need protein; after water, it is the second most abundant substance in the human body. Basically, protein is a compound of amino acids, and it takes 20 kinds for the body to put together a "complete" protein. The body can produce roughly half, and the rest must come from food (accordingly, they're called essential amino acids), pretty much on a daily basis. (This also means you don't need to overeat protein, since your body disposes of what it doesn't need anyway.) The most convenient source of complete protein is animal foods, but there are some complete vegetable sources, and many nearly complete sources that complement each other.

It might be a stretch to say that protein is overrated, but it isn't a stretch to say that you don't need to worry about it much. *Any reasonably balanced diet that you devise, any diet that contains a minimum of junk food and refined carbs, is going to give you enough protein.*

Whether the importance of protein is overstated or not, almost no one would dispute that the vast, overwhelming majority of Americans get more protein than they need, and that almost all the excess comes from animal products. This isn't surprising: We grew up eating meat, most of us like it, and meat is quite high in protein. Meat satisfies us culturally and by its flavor and texture, and if you're sold on our need for a lot of protein, it works in that sense, too. No wonder so many people argue that we need it.

Americans consume 10 TIMES as much meat as people in many developing countries.

We even eat too much protein by our own government's generous standards.

But remember that we eat twice as much meat as the world average, and 10 times as much as people in many developing countries. Though it's historically accurate to say that just about all cultures have maximized their meat consumption, it's equally true that people thrive with adequate calories but not a lot of animal protein. If the American high-protein diet were the ideal, you might expect us to live longer than countries where meat consumption is more moderate. But as I noted earlier, that isn't the case; we're second-to-last in longevity among industrial nations.

We even eat too much protein by our own government's generous standards. The recommendation is one-third of a gram per pound of body weight, so if you weigh 150 pounds you should be eating 50 grams a day, according to the USDA. Most of us exceed the RDA by 30 percent or more, and some experts believe that the RDA is already too high.

We do need protein, and athletes and bodybuilders need more than the rest of us. But one-third gram per pound of body weight is plenty. More than that causes calcium loss (though some people eat high-protein dairy specifically in order to increase their calcium intake), increases your need for fluids, and makes your kidneys work harder. And some recent research indicates that protein is related to the immune malfunction that causes food allergies.

There's some evidence that vegetable protein is more beneficial than animal protein.

So it may be that instead of worrying about not getting enough protein, you should avoid eating too much. If you stop using protein as an excuse to eat animal products, and if you replace animal products with plants, your body will benefit in several ways: you'll be eating more micronutrients, more fiber, and healthier fats. There's also some evidence that vegetable protein itself is more beneficial than animal protein.

Carbohydrates

Carbohydrates are sugar molecules found in most foods, especially sugars, starches, and fibers. They are divided into two categories—simple and complex—but the way your body treats the many possible chemical combinations is more complicated than that.

Simple carbohydrates are made from the simplest sugars either alone, as in sugar and other sweeteners, or combined to form the simple starches that are found in refined foods, like pasta and bread made from white flour. Multiple simple carbohydrates are known as complex carbohydrates, which are abundant in whole grains and vegetables.

Carbs themselves are not the problem, just as protein and fat aren't problems. Just as some protein sources are better for you than others, there are "good" and "bad" carbohydrates. Almost every nutritional expert agrees that simple carbohydrates are at best useless calories and at worst damaging, at least in the quantities in which we consume them. They serve almost no nutritional purpose besides getting calories into your body, something that is not a challenge for all but the most impoverished Americans.

Instead of worrying about not getting enough protein, you should avoid eating too much.

So if you eat anything approaching a typical American diet, you should undoubtedly eat *more* carbs, but complex

Simple carbohydrates are at best useless calories and at worst damaging.

ones—legumes, whole grains, and real whole grain breads. (I say "real" because most supermarket breads that are labeled "whole grain" are a hoax, containing, for example, 20 percent whole wheat flour and 80 percent white.)

The problem with most diets, whether low-fat or low-carb, is that in the long run they tend to raise the number of calories you eat. In fact, the low-fat craze caused millions, maybe tens of millions, of Americans actually to gain weight, because they were reaching for "low-fat" but high-calorie carbs. When you drastically reduce carbs—almost all carbs, as some radical diets like Atkins recommend—most people never quite feel satisfied, no matter how much meat they eat. So they end up eating carbs and throwing off the precarious body chemistry that allows a low-carb diet to work in the first place.

Your body can scarcely tell the difference between white flour and white sugar. Either, in excess, will increase the possibility of your gaining weight and developing type 2 diabetes. And there are simple carbohydrates that are even more damaging, especially fructose.

The special case of simple sugars

Let's talk about corn. You probably eat about a dozen ears of corn a year, yet agribusiness produces over 9 billion *bushels* a year. Much of that is fed to livestock, but much becomes high fructose corn syrup (HFCS), a manufactured sugar that has replaced cane and beet sugars (sucrose) as the primary sweetener for many kinds of foods, from sodas to savory items. Food manufacturers prefer HFCS because it's cheap, it's easy to use, and it increases the shelf life of processed foods.

Your body can scarcely tell the difference between white flour and white sugar.

But HFCS creates many problems. If you eat too much sugar of *any* kind, the liver converts it to fat; but large amounts of fructose (like that contained in soda) seem to stimulate hepatic lipogenesis, the liver's ability to make fat.

And if you eat too much fructose, the liver becomes even better at doing so.

Worse, it appears that the more fructose you eat, the hungrier you feel.

Furthermore, it appears that too much fructose and glucose in the diet may disable the body's ability to regulate testosterone and estrogen levels. That disability is associated with an increased risk of acne (teenagers' fears about sugar are correct), infertility, ovarian cysts, cardiovascular disease, and uterine cancer in overweight women.

Put simply, if you eat a lot of sugar (or simple carbs in general), you had better eat a lot less of everything else, or you're going to gain weight. This is especially true of sugar in the form of soda sweetened with HFCS (as most soda is), because these calories do not fill you up in the same way as calories you get by eating, even by eating sugar. Sadly as a nation, we get an astonishing 7 percent of our calories from soda. (One experiment compared soda and jelly beans; you're better off with jelly beans.)

Each American eats an average of 1 CUP of sugar a day.

It appears that the more fructose you eat, the hungrier you feel.

Despite all this, we're not eating less sugar; we're eating more. As a nation, we now produce about 80 pounds a year per person of corn-based sweeteners (mostly HFCS), an increase of about 16 pounds a year since 1985. Over that same period, per capita sugar production has remained virtually the same: about 63 pounds. Not all of this is actually eaten, but a good estimate is that per capita consumption of sugars is at least 125 pounds of sweeteners a year, or about 5 ounces a day: about 1 cup, or 600 calories.

The right carbohydrates

Whole grains are a different story. As it turns out, the parts of the grain that are removed to make white flour, white rice, and so on, are exactly the parts you want to be eating. Contained in whole grains and seeds are micronutrients that are not found in white carbohydrates—micronutrients whose roles are not yet well defined but which appear to be beneficial—as well as a lot more fiber.

Fiber is the category of carbohydrates that your body doesn't digest. Though it provides no direct nutrition, fiber is believed to reduce the risk of heart disease, type 2 diabetes, constipation, and other digestive disorders. It's found in all plants, and generally falls into two types: soluble fiber (that which dissolves in water) and insoluble fiber (that which does not).

Most high-fiber foods have some of both types. To prevent disease, it appears that soluble fiber—which is prevalent in some whole grains (like oats or barley, though not wheat), some legumes (like soybeans and kidney beans), and citrus fruits—may clear the body of fat and regulate the way sugars are burned and stored. It also helps make you feel full and satisfied after eating. For relief of constipation, insoluble fiber (high in most vegetables and fruits, most legumes, and most nuts and seeds) is better. There's no reason not to eat both.

Most Americans get only
15
grams
a day of dietary fiber—half the recommended amount.

In fact, most Americans are getting only 15 grams of fiber a day, half the recommended amount. But if you change your diet to emphasize fresh fruits and vegetables, whole grains, and nuts—in other words, if you eat as though food matters—you'll easily bump your daily fiber up to the recommended 30 grams.

The relationship between refined carbohydrates and type 2 diabetes and pre-diabetes (simply high blood sugar, which may lead to diabetes) is a rare area of almost universal scientific agreement. My own pre-diabetic condition essentially went away when I changed my diet to exclude most simple carbohydrates.

Again, this part of the "plan" is simple: instead of eating white flour (this means most commercially available bread, bagels, cake, muffins, pizza, sandwiches, and so on), sugar, processed foods (including many boxed breakfast cereals), or pasta, I eat whole grains—oatmeal, cornmeal (polenta or grits), rice, wheat, quinoa, barley, and some whole grain breads. (But they must be *real* whole grain breads, not those made with 20 percent whole wheat flour.) By nighttime, I'm really ready for some crusty white bread or cookies, but they make up a very small part of my caloric intake.

Once you limit processed foods, refined carbs, and animal products, fat becomes nearly a nonissue.

Fat

Fat has become a national obsession. Not only how much fat but what kind of fat we should eat is endlessly debated. But you don't have to participate in this debate, at least not much: once you limit or avoid processed foods, refined carbs, and animal protein, fat becomes nearly a nonissue, even if your major goal is to lose weight.

You actually need quite a bit of fat to live. (Your brain is approximately two-thirds fat, and this fat has to come from somewhere.) For most of human history, it's been among the hardest nutrients to get enough of. Only when you eat far too much of the wrong kind is fat a problem, and sane diets avoid that problem quite effortlessly.

For a time, there was a near consensus that saturated fat was unhealthy, and that eating too much of it led to heart disease and other diseases. But most foods contain at least some fat, and all naturally occurring or easily produced fats have a role in a healthy diet.

Recent research indicates that the most crucial factor in heart health is the balance of fats in your bloodstream. The right balance means that you're not adding to the cholesterol your body is already producing, and that you are mitigating any additions with monounsaturated fats and omega-3 fatty acids, like the ones found in oily fish and, to a lesser degree, nuts. But that balance is skewed in the wrong direction by the typical American diet; we get too much of the fat that occurs in animals, and not enough of the kind that occurs in plants.

It's time for a word about cholesterol, not because cholesterol is so important but because the anticholesterol campaign has been so visible for so long that it still concerns people. As with everything about nutrition, cholesterol has turned out to be far more complicated than was once thought. Like fats in general, some cholesterol is "good" for you, and some is "bad" for you. The ratio between the two kinds is probably more important than the total amount. Equally important is that the amount of cholesterol you eat is much less likely to influence the total cholesterol in your blood than the amount of cholesterol produced by your liver.

What determines how much cholesterol your liver makes? Not the cholesterol you eat but the kind of fat you eat. Monounsaturated and polyunsaturated fats tend to raise the good type of cholesterol while lowering the bad. Saturated fat, found most in animals, tends to be more or less neutral—not so bad, in small quantities at least—raising both types of cholesterol equally. Trans fat—the stuff manufactured to produce margarine, solid shortening, and much of the fat that goes into processed and junk foods—raises the bad while lowering the good. (So much for 30 years of advice about eating margarine!)

There is nothing radical in what I'm recommending, and it involves the principles behind *Food Matters*. If you eat naturally occurring fats—those found in or derived readily from plants and animals, and you eat less of the animal-based fats, your diet will be a better one. You don't have to give it much more thought than that.

Eating like food matters

The evidence overwhelmingly supports a more traditional diet—what I'm calling sane eating—in place of the modern American diet. Mediterranean, Indian, Middle Eastern, North African, French, and most traditional Asian diets all contain far fewer animal products and refined carbohydrates than ours. Base your preferred diet on any traditional eating style you like; the point is that once you get into the habit of eating sanely, it becomes second nature. That isn't surprising, because it's far more natural than eating processed food, junk food, and historically unprecedented amounts of (badly produced) animal products, none of which existed for 99 percent of human history.

Gorge on plants. Literally.

You will do yourself a favor every time you eat a vegetable in place of anything else.

Let's look at the general principles of the style of eating I'm advocating:

Eat fewer animal products than average. Say, an average of 1 pound of meat, or at most 2 pounds, each week, or a small serving daily. (If some of these servings are fish, so much the better.) Eat correspondingly small amounts of eggs and dairy foods, and think of all these things as treats, not staples. Milk in your cereal or cream in your coffee isn't going to make much difference, though alternative milks from plant foods—like soymilk, oat milk, and nut milks—can be decent substitutes. Remember, this is not about deprivation or ironclad rules, but about being sensible.

Eat all the plants you can manage. Literally. Gorge on them. Salads, cooked vegetables, raw vegetables, whole fruits—cooked or raw or even, in moderation, dried. There are hardly any limits here (though you don't want a diet based entirely on starchy vegetables like potatoes). I might say that green, leafy vegetables are probably the most beneficial of all these foods, but you are going to be doing yourself a favor every time you eat a vegetable in place of anything else, so don't worry about it.

Make legumes part of your life. Whenever you eat beans instead of an animal product, everyone wins. Especially if you're concerned about protein (again, I don't think you need be), eat legumes daily.

Whole grains beat refined carbs. You shouldn't eat "unlimited" amounts of grains, as you would other plants, but eating grains several times a day is fine. You might have whole grain cereal or bread at breakfast, whole grain bread or a grain dish at lunch, popcorn for a snack, a grain dish at dinner. In any case, eat far fewer refined carbohydrates; they are all treats, not off limits but to be eaten only occasionally (and with gusto).

Snack on nuts or olives. These are something of a special case, because they're high in calories. But you're going to be eating so many fewer calories that you can afford to eat a couple of handfuls a day. I make my own trail mix and eat it along with some fruit almost every afternoon at work.

When it comes to fats, embrace olive oil. That's where you start. You can use butter when its flavor or luxury is really going to matter to you. Use peanut oil or grape seed oil for stir-frying (or any frying), use dark sesame or nut oil for extra flavor, and you really don't need much else. (I'm not a fan of canola oil, but use it if you must.) Don't worry too much about quantity. Don't start drinking oil, or eating fried food daily; but using oil for dressing or cooking is not a big deal, provided you're not eating many refined carbohydrates or animal products.

Everything else is a treat, and you can have treats daily. Listen to your body: Are you losing weight, feeling fine, getting results that make you and your doctor happy? Keep it up. Are you not getting the results you want? Cut back on treats, and eat more plants. Treats include alcohol (a lot of useless calories and carbs come in the form of wine and other alcoholic beverages), snack food, refined carbs (including good, crusty, artisanally made bread), and sweets of all kinds.

Within these general guidelines, eating like food matters is extremely flexible. You can try some of the techniques that work for me (page 73), or just eat more sanely at every meal, then snack and allow yourself small indulgences throughout the day. If you eat moderately and always try to put as many plants on the plate as possible, you'll be in the ballpark.

You might start by eating 10 percent less meat, less refined carbs, and less junk, and replace that food with plants, but I think 25 percent is probably a better starting place, and one that will show you results more convincingly. (Frankly, I didn't find it very hard to cut junk food out virtually altogether. Meat and carbs are a little more difficult, but remember that you're not going to give them up entirely.)

Another strategy is to load up your plate with salad, vegetables, and whole grains, and then put some meat, fish, or poultry on it as well. Better still, eat that big plateful of plants first, then go back for a small piece of meat. This is a very "Italian" style of eating.

Rely on meat for its flavor, not its heft.

In the morning, for example, you might eat a couple of spoonfuls of yogurt with a big bowl of fresh fruit and a sprinkle of real muesli or granola. For lunch, have half a tuna sandwich on real whole grain bread with a big salad or vegetable soup. For dinner you and your friends and family go out and share two entrées and load up on vegetable side dishes and appetizers, then order one dessert with four spoons.

Not everyone responds to making changes the same, somewhat drastic way I've done. You can transition into this slowly, taking baby steps toward whatever goals you set for yourself. Some suggestions follow.

Cut back on animal protein gradually. Rely on meat for its flavor, not its heft, using more vegetables in your favorite meat dishes. Make pork and beans with half the meat (you'd be amazed at the flavoring prowess of just one sausage) and add extra beans or vegetables or both to the pot.

If you're having company, you might roast a chicken (not two), along with a load of root veggies, and a couple of other vegetables dishes or salads. Next time you grill burgers, make the patties smaller, and toss eggplant, onions, potatoes, summer or winter squash, and portobello mushrooms on the grill too. Or try Meat-and-Grain Loaves, Burgers, and Balls on page 278, which combine ground meat with grains.

It's the same with dairy foods. Add a couple of big slices of tomato and some thinly sliced pickles and onions to your next grilled cheese sandwich and cut back on the cheese. Start a batch of scrambled eggs by sautéing mushrooms or greens in the pan and try adding in one egg instead of two (or check out the frittata recipe on page 170). Blend a smoothie (page 163), using frozen fruit and just enough yogurt or milk to give it some body. And again, give nondairy milks a try; you might like them.

Eat Whole Grains with Other Foods. Experiment with uncommon grains like millet or quinoa by stirring a couple of spoonfuls into a stew or soup as it cooks; toss some cooked grains into a salad or a stir-fry at the last minute. Or just play around with new grains—barley makes a great "risotto"—they're easy enough to like. Try making your own bread (pages 154, 156, and 224).

Depend on Seasonings. Good fruits and vegetables rarely need more than a sprinkle of salt, but if you're feeling hungry for more variety, try different herbs and spices, alone or in blends. Some people find that using seasonings they associate with meat—like soy sauce, pesto, or chili powder—is a good way to make the transition to enjoying plant foods.

Always Carry Snacks. This is important, since fast food is everywhere and taking a couple of minutes before you head out will make impulsive stops for junk food less tempting.

Dried or fresh fruit and nuts are the easiest options, but with a little planning and a small cooler or thermos, you can travel with hummus and crackers, cut-up vegetables, a container of excellent juice, some olives, a peanut butter sandwich, a bag of granola, a cup of soup, or some fresh popcorn.

Sane shopping

Shopping for sane eating is easy. Even if you shop for tonight's dinner on the way home from work, you'll have no trouble pulling something together. This will be increasingly true as you shift from animal products to plants, which generally cook quickly.

You can shop pretty much anywhere. Supermarkets and so-called natural foods stores have plenty of whole grain foods and produce; farmers' markets are often your best choice for vegetables; and international stores often offer variety that you're not going to find elsewhere. You don't need any special food or ingredients to cook sanely, though obviously the more variety you bring to the way you eat, the more you'll enjoy eating.

Make sure your refrigerator is full, mostly with fruits and vegetables.

In a nutshell: Buy lots of fresh and supplement with some frozen and dried produce. Buy correspondingly less meat, fish, and poultry, but buy the highest quality you can afford, ideally from sources you know and trust. Stay away from any processed food that has more than five ingredients; and ingredients with more than three syllables (in other words, stay away from preservatives and additives).

Let's tackle these food categories one at a time.

Produce is the most important, and I can't stress this enough. Make sure your refrigerator is full at all times, mostly with fruits and vegetables. (They take up a lot of room; that's why they have low caloric density.) Keep bowls of fruit (vegetables, too) on the kitchen counter or dining room table—they're gorgeous, after all, and if you live with them you'll eat them. Along with the most perishable types, be sure to stock carrots, potatoes, root vegetables, winter squash, citrus, apples, and cabbage.

I keep some frozen vegetables on hand; my favorites are peas, "fresh" shell beans, Brussels sprouts, and corn. I also stock canned beans and tomatoes, and sometimes prewashed bagged greens and even cut-up salad bar veggies—anything that makes it easier to eat in my new style.

It's worth thinking about the amount of packaging and processing involved in your food; try to buy food in bulk, and bring your own bags (you probably know all this). But at the risk of being repetitive, let me remind you that cutting back on animal protein is among the most important environmental contributions you can make, at least when it comes to food.

The same common sense applies when you buy meat, poultry, eggs, and dairy. If you're concerned about animal welfare and want to avoid hormones and antibiotics in your meat, then you're either going to have to buy organic food (that's the only label distinction that is even remotely regulated; see page 101 for more) or purchase animal products from a place you know and trust.

Fish is a special case. Wild fish, obviously, is organic, though there are concerns about mercury and heavy metals in tuna and swordfish. But much of it is also endangered, so it sometimes can't be purchased with a good conscience. Farmed fish often has many of the same problems as farmed land animals, including the use of antibiotics, environmental damage, and insipid taste.

On the other hand, fish can be the healthiest animal product you can eat. It contains few harmful fats and often high amounts of omega-3s.

This doesn't mean you should start eating fish seven days a week; it's still an animal product, and there are still many good reasons to limit your consumption. But if you can find fresh (or well frozen) wild fish that's not on any endangered species list (the Monterey Bay Aquarium's Web site mbayaq.org has a list, broken down by region), and isn't on any warning lists (yellowfin tuna, for example, is not currently endangered though it does contain high mercury levels), and you don't object to it for ethical reasons, it's probably the best choice in animal foods.

By supporting an alternative to the "industrial meat complex," you're rejecting that type of agriculture in favor of something far better for the planet and for you.

In general, wild fish and well-raised forms of animal protein are going to cost you more, and sometimes a lot more, than their conventional counterparts. But in general, rebalancing your consumption to achieve a plant-centered diet will probably reduce your overall grocery bill. You can enjoy the savings or use it to upgrade the products you buy. A $20 a pound price tag for a couple of servings of fish or meat is unquestionably high, but if you're eating only 1 or 2 pounds a week, if you're a typical American it's probably not more than you're spending now.

And each time you make a decision to support an alternative to the industrial meat complex, you're rejecting that type of agriculture in favor of something far better for the planet, and for you. Change will come, and "conventionally" raised meat, fish, poultry, and dairy foods may become more acceptable.

The five-ingredient rule

To eat sanely, you don't need to know how to read everything on a label, though it's easy enough. It's not as easy, though, as this rule (originally "mandated" by Michael Pollan and others): avoid anything with more than five familiar-sounding ingredients.

Before going further, it's worth mentioning that, applied strictly, this would eliminate conventionally raised meat from your diet, if it were labeled. Because if you listed the ingredients that went into producing it, the label might include alfalfa cubes, barley silage, dried cattle manure, blood meal, coffee grounds, chicken fat, corn and cob meal, ammonium sulfate (for fertilizer), hydrolyzed feather meal, ground limestone, cooked municipal garbage, linseed meal solvent, oat straw, potato waste, dried poultry manure, soybeans, wheat, antibiotics, and any pesticides or herbicides used in the corn and soybean fields, just to name a few.

Meat isn't labeled but most packaged food is, and though the five-ingredient rule won't eliminate all junk food from your diet, it will go a long way toward eliminating junk food, and it will simplify your shopping.

Of course there are levels of "junk": there are potato chips made with two ingredients (potatoes and oil) and ice creams made with only four or five. These, of course, fall into the category of treats. But it's the chips product, and faux-fat ice cream, and frozen dinners, and all the other stuff made with 15 or 20 ingredients that you should pass up altogether, and forever: there is nothing good about them, even in limited quantities.

There are a few ingredients that I try not to eat even if the product otherwise passes muster. These include hydrogenated anything, monosodium glutamate (sometimes hidden behind terms like "natural flavorings" or "spices"), high fructose corn syrup, and anything I've never seen—which includes about 80 percent of the ingredients on junk food labels. (Pasteurized processed cheese product? Guar gum? Silicon dioxide?)

If the package, jar, or box in your hand passes the five-ingredient rule, and you still want to read the label for calories, protein, fiber, and so on, more power to you. But as long as you're eating plenty of fruits, vegetables, legumes, and nuts, and a small amount of meat, fish, and dairy food, you'll be in fine shape nutrient-wise.

On buying organic, or local, or sustainable, or whatever

It's a personal choice, but if you decide to steer clear of conventionally raised meat, the logical next step is to choose organic, local, and sustainably raised foods. This is especially true if you're trying to minimize your impact on the environment. But within each of these distinctions is a range of practices, some regulated by the government, others not.

Legally, in order for a food to be labeled "organic" it must be certified by approved third-party companies. The USDA has issued pages and pages of rules and regulations, but the most important are that

I don't believe "eating organic" is essential to sane eating.

- No nonorganic pesticides or fertilizers were used.
- In the case of animals, no antibiotics or hormones were used.
- No genetically modified crops were used at any stage of production.

This is fairly clear, although the benefits remain debatable—that is why I don't believe "eating organic" is essential to sane eating.

By contrast, "natural," "cage-free," and even "grass-fed" are not nearly as well defined or regulated. So unless a label makes statements like "no pesticides" or "no antibiotics," you have no idea whether or not they were really used. (A true skeptic doesn't believe such declarations anyway, and I sometimes fall into that camp. But what are you going to do?)

Certified organic products can display the USDA logo as a seal of approval. For single-ingredient foods like produce, pantry, dairy items, eggs, and meats, approval means that the product complies with the USDA standards. For prepared and processed foods, those made with more than one

Can a head of lettuce that travels 3,000 miles by truck still qualify as "organic"?

ingredient, the certification system allows for a range of labeling options, from 100 percent (meaning that all the ingredients used to make the product were organic) to "made with organic ingredients" (at least 70 percent). If the food is labeled simply "organic," along with the shield, it means that between 95 percent and 100 percent of the ingredients are certified organic.

This is not ideal, but it's the only regulated distinction from conventionally raised food we have. Since all this certification is expensive and onerous, you should expect companies and people who have really earned it to display it. (The fine for selling something uncertified as organic is $11,000; a deterrent for very small producers but not at all for larger ones.) At farmers' markets or other places where labeling is more ambiguous, you're going to have to ask, and then use your best judgment.

Organic food, of course, has become big business; with Big Food companies continually snapping up organic companies and creating new organic products, this is among the fastest-growing sectors of the food industry. That raises questions about mass production, mass pollution, and mass distribution—the same issues that are raised about conventionally produced food. To me (and to a lot of other people), all this defeats the purpose, which is to produce food in a way that sustains us and the planet. Can a head of lettuce that travels 3,000 miles by truck, or a piece of fish that's been flown halfway across the world, still qualify as "organic"?

Enter locally raised food, the trend that led to the term "locavore." A locavore is someone who eats food that's grown locally, usually within a couple of hundred miles. The environmental benefit here is that the food doesn't travel far. But there's the inconvenient fact that if you live anywhere in the northern half of the country you are not going to have a lot of options come winter. (Some people will argue this is the direction in which we are and indeed should be heading, but it's a tough sell in twenty-first-century America.) And sometimes economies of scale and distribution may make it not only economically but environmentally "cheaper" to ship products from afar.

Still, eating locally has many more positives than negatives. It makes people think about seasonality, and about how ridiculous it is to eat summer fruits and vegetables in January; ending or limiting that habit would be unquestionably good for the environment (and for the cook; there is joy in eating seasonally).

Locavorism has also been a boon for urban and suburban farmers' markets, and—more important—for the farmers who supply them. That in turn helps rural economies. Most important, produce fresh from the ground and animal products raised by real humans provide the most nourishment and the most enjoyment.

Local food is usually expensive, and it can't provide everything for everyone, but it's sensible and as environmentally sound as it's going to get. I'd choose nonorganic conscientiously raised local food over corporate organic food in just about every case.

Eating locally has many more positives than negatives.

This is especially so because many small farmers skip the cost of getting organically certified but practice sustainable farming, which is often a combination of the best of organic with the best of local. "Sustainable" is a fuzzy term, and it's not legally defined, but the idea is to produce food in a system (we used to call this a "farm") that uses modern versions of old agricultural practices and a minimum of artificial inputs. In other words, you may have some livestock in a barn, and you collect the hay used for bedding (rich in animal wastes, naturally) and compost it for use in the fields. Chickens roam freely, pecking for bugs and contributing their own fertilizer to the soil. And of course you harvest eggs, milk, and meat.

Is there enough land? Are there enough knowledgeable farmers? Is there a localized distribution system that can

Shop where you like and buy what looks freshest and most appealing to you.

You don't have to cook to eat sanely—but it helps.

support truly sustainable agriculture? I don't think so. As you know from earlier discussions, industrial farming was designed to streamline agriculture and resolve these questions, so my guess is that modern sustainability is going to mean a food production and distribution system that looks different from the way it does now, but it's not going to look like an updated version of the early twentieth century.

The bottom line is this: shop where you like and buy what looks freshest and most appealing to you. Eating as though food matters will heighten your awareness of these complicated issues, and you'll naturally pay more attention to the impact your choices have on you and the environment. Are your choices going to be perfect and free of hypocrisy? I'd be lying if I said mine were. But I can't repeat it enough: the aggregate of even the smallest changes equals big changes.

The *Food Matters* kitchen

You don't have to cook to eat sanely, but cooking helps, and cooking is what Part II of *Food Matters* is about. If you already know how to cook, you have a head start. If you don't, let me tell you that cooking can be a joy. Over the years I've developed some techniques that help minimize the work and maximize the yield. Here, then, are the basics of the *Food Matters* kitchen.

Planning—and cooking—to eat more plants

People never think they have time to plan, but I'm going to make the case that planning (call it thinking ahead if you like) actually saves you time. Start with this premise: You're going to eat what's handy. (This is how junk food manufacturers get rich.) If you always have veggies ready for cooking, quick assembly into impromptu dishes, snacking, or taking on the run, that's what you'll eat. Here, then, are some simple suggestions.

When you cook at home, wash and prepare vegetables in bulk. Most cut-up veggies, properly stored, will keep for days, with no problem. It takes far less than twice as much time to prep twice as many vegetables: the equipment is out, the water is in the sink, and your attitude is adjusted. Peel a bag of carrots instead of a couple, and put the extra in water in the fridge. Ditto celery. Clean a head of cabbage and cut it into wedges. Wash a couple heads of broccoli or cauliflower and trim them into florets. Spend some time each day doing this sort of stuff (you can talk on the phone at the same time, though sending e-mail is tough), and you'll always maintain a week's worth of fresh vegetables, ready to go.

This is especially true for salad greens; it takes hardly any more time to wash a head of lettuce, or two heads for that matter, than to wash a few leaves. Buy a salad spinner, the kind without holes in the bottom for the water to drain out. This is a the best way to wash (and store) all kinds of greens (and other veggies). A spinner with a tight seal can extend the life of veggies for up to a week; leave a few drops of water in the bottom to help retain moisture.

It takes far less than twice as much time to prep twice as many vegetables.

Similarly, cook vegetables, legumes, and grains in large quantities. Here's a case where you also conserve energy, since it takes just as much energy to roast or grill a pound of vegetables as to cook three pounds. And if you have the oven on to roast a chicken, why not add a pan of vegetables to the rack below? Similarly, you can cook three cups of rice or oatmeal as easily as you cook one cup, or a few pounds of potatoes as easily as one (very handy when you want potato salad). It's the same effort, and cooked vegetables and grains are easy to store and reheat (page 136).

I'm a big fan of leftover finished dishes, but if you set aside plainly cooked extra vegetables, they're even easier to vary and reuse. And, every week, cook a full pot of beans and at least one pot of whole grains. This stuff keeps really well and has infinite uses.

Precooking is the best way to extend the life of food that is threatening to go bad on you. Chop up a bag of apples, skin on if you like, for a batch of quick applesauce. Make a compote out of assorted fruit that seems a little over the top.

And use your freezer. It's running all the time anyway, right? In fact it'll run more efficiently full than empty. (You're not buying microwave dinners anymore, either.) Fill it with staples, as you would a pantry: frozen vegetables, precooked beans, leftovers, grains and nuts for long-term storage, and so on.

Eating in restaurants or taking out

Eating sanely is easiest when you're home: you have control over what's in the house, what you prepare, and what you put on the table. If you stayed at home all the time, you could determine just about everything that went into your mouth.

No one does that. We go to work; we travel; we eat out for pleasure. And it's outside the home that things become tricky.

This, I think, is where the *Food Matters* strategy really shines, because among its chief principles is to let yourself go. "Cheating" (it isn't really cheating; it's part of the plan) is not only allowed but encouraged. No one wants to give up pleasure if they don't have to, and I'm not asking you to. If you eat lunch or dinner out, and you don't want to maintain the general sane eating habits—well, don't. As I've said, what works for me is to be ultrastrict from dawn until dusk, and then let myself go more or less wild, although the wildness has become more moderate as my habits have changed.

So that's the first rule: don't let yourself feel too guilty.

In the *Food Matters* strategy, cheating is not only allowed but encouraged.

Having said that, I know full well the temptations of lunches at work and meals grabbed on the run. And you can't eat sanely unless you can be disciplined most of the time.

There are two basic strategies. One, carry your own food. Two, figure out in advance what you're going to eat. If you carry your own food, you're always safe. At my work desk, I have popcorn (and a little covered-bowl setup that lets me pop it, with or without oil, in the office microwave); various bags of nuts and dried fruit, sunflower seeds, and the like; a few pieces of fruit, which I replenish once or twice a week from home or a nearby vegetable stand; and sometimes some whole grain crackers.

Needless to say, there are many days when this assortment doesn't cut it and I head upstairs to the company cafeteria for lunch. That's easy enough: I look for vegetable side dishes (quite cheap, by the way), salads of any and all types, beans, olives, grilled vegetables, and, if the pickings are slim, raw broccoli, cauliflower, tomatoes, cucumbers, carrots. I sometimes fill two plates with this stuff, and I assure you that while the volume is high the caloric density approaches zero.

The hard part is lunch out. I can make the easy decision, which is essentially to postpone discipline, and allow my one big meal of the day to be lunch. But I've learned that in reality—again, I'm me, and you may be different—this is a slippery slope; when I eat a big lunch, I still want a glass of wine later with my light dinner, and the glass of wine often leads to another, and to "a little pasta," and suddenly my light dinner isn't so light.

So I try to steer business associates to lunch places where I know I can do pretty well, those that have lots of vegetarian entrées (Indian restaurants are always a good bet), or where I can get a salad, some grilled vegetables, and maybe a piece of fish. Takeout is along the same lines. It's the same sort of intuitive planning ahead that I discussed earlier in this section, but now someone else is cooking.

On the road, in airports, in strange cities, things are a little more difficult. Sometimes I order two salads, or salad and soup and a side of vegetables. Sometimes I tough it out and buy nuts, carrot sticks, whatever I can find, and figure I'm going to be a little hungry that afternoon.

And sometimes I give up. This is a long-range plan, after all, and what happens on any given day matters not at all. Overeating, or eating "badly," in the modern American style, is not a physical addiction, like smoking, drinking, or heroin; you can recover from an off day the next day, with no lingering ill effects, even psychological ones.

For a more detailed look at a typical day of eating, head to Food Matters Meal Plans.

FOOD MATTERS PART II

How to Cook Like Food Matters

Eating like food matters is easy. You don't need to count calories, weigh portions, or find unusual ingredients. In fact, for the most part the ingredients are absolutely regular.

But this is not to say that the recipes are the same ones you find everywhere else. Many of them have what might be considered unusual constructions; and as a group, they reflect the general principles outlined in the preceding chapters of the book: less meat, more plants. This makes them more traditional than the vast majority of recipes printed in most twentieth-century cookbooks, which tend to represent a kind of excess that is no longer considered either desirable or practical.

Not all the *Food Matters* recipes are vegetarian. Some are, but others are what you might call flexitarian: meat or fish is optional. Those that may seem most unusual feature animal products in smaller proportions than is traditional in the United States. So when you grill meat and vegetables to make kebabs (page 246), you'll be grilling more vegetables than meat; similarly, when you combine meat and beans to make cassoulet (page 262), you won't be using three pounds of meat to serve six people; and even chicken breasts are stuffed with vegetables (page 268).

The general idea is that whether you use these recipes individually, occasionally, or religiously, you'll shift the balance of your calorie intake from predominantly animal products to predominantly plant products. You can use the knowledge you gained from reading the preceding chapters, or refer to the meal plans on pages 122–29, to put together dishes, meals, or daily schedules that appeal to you.

None of this assumes that you're going to stop eating "normal" food; in fact, the recipes here are "normal" food. I do assume, however, that if you want a grilled steak or a bowl of ice cream you won't be looking here to find it. Again, both steak and ice cream fit perfectly well into a Food Matters pattern of eating, but with far less frequency than most of us have become used to. And to the extent that the recipes here represent the bulk of your diet, you'll probably be eating far better than you ever have.

Since the recipes here, or recipes like them, will form the backbone of your diet when you start eating well, it makes sense to have your pantry well stocked with the basics, and to begin intentionally overcooking. Many staples of the Food Matters diet can be cooked ahead of time and frozen or refrigerated, to be used in the coming days or weeks, so it's often worth your time to double or triple a given recipe and tuck it away. (You'll find the recipes I'm talking about in the section called "The Basics," beginning on page 131.)

For example, there is almost no reason to cook less than a pound of dried beans or whole grains: A large amount takes the same time as a small amount, and both foods keep well, refrigerated, for up to a week, and frozen indefinitely. You'll probably wind up eating beans several times a week anyway, and grains are easily heated up (or eaten cold) for breakfast cereal.

Again, if the recipes here (and the general style of these recipes) make up the bulk of your diet, you'll probably be eating far better than you have ever eaten before. To cook in the Food Matters style you'll probably want to adjust—to a greater or lesser degree—the foods you stock regularly in your cupboards, fridge, and freezer—in short, your pantry.

Some are obvious, and you already have them (olive oil, for

example); some are not so obvious. There are some that are complete musts, whereas others are more or less optional, though they can help you in the long run.

So, I've created two lists: The Essential Food Matters Pantry and The Advanced Food Matters Pantry. If you maintain a well-stocked pantry, you'll be able to produce most of the recipes here without doing much (if any) other shopping. And since most of these ingredients keep nearly indefinitely, it's worth a little initial investment in time and money to get them into the house. Buy what tastes good to you; price isn't always the best indicator, but you want these ingredients to be high-quality.

The Essential Food Matters Pantry

Just about everything on this list belongs in every kitchen pretty much all the time.

Grains. These are mostly whole, and include rice, cornmeal, and whole grain flours. Buy an assortment, preferably in bulk from a store with a fair amount of turnover. But don't buy grains by the ton, at least at first; you will have preferences as you get to know them. Store a pound or so in the cupboard, with excess in the freezer or fridge.

Beans. Like grains, buy an assortment of dried beans. But unlike grains, legumes come in thousands of varieties, ranging from white to black, from yellow to red to celadon, from tiny to huge, from round to kidney-shaped. Again, buy a pound or so at a time, and don't buy more than you're going to use in the next few months. Canned beans are fine, too, but they're more expensive, the selection isn't as great, and they don't taste as good. Frozen beans are an excellent convenience food, though sometimes anything other than lima beans are hard to find.

Olive oil. Your go-to fat. Extra-virgin, in almost every case. Country of origin doesn't matter much. Price does, but the $10-a-liter stuff is perfectly fine for most uses. Store a pint or so on the counter, the rest in a dark cupboard or the fridge; it keeps for only a few months. (You can extend its life by refrigerating the portion that you're not going to use in the next couple of weeks.)

Other oils. You'll need something more neutral for cooking Asian-style dishes or for pan-frying at high heat. In the recipes, I usually suggest either peanut oil or other vegetable oils. I use grape seed (though it can be hard to find), since I'm not a fan of canola (it's sticky and, to me, has an off flavor). Other vegetable oils, like sunflower or peanut, might work well for you. The key is always to buy minimally processed, high-quality oil, cold-pressed whenever possible.

Staple vegetables and fruits. These range from much-used seasonings, like onions and garlic, which you should always have on hand (shallots are nice, too); to frozen vegetables like spinach, peas, and corn, which it pays to stock; to fresh vegetables, which you have to purchase at least a couple of times a week. Keepers like carrots, potatoes and sweet potatoes, celery, lemons, and limes, can be replenished as you use them.

Fresh herbs. Something as simple as parsley can make all the difference; and almost all herbs, especially basil, mint, dill, rosemary, thyme, and cilantro, are great to have around.

Spices. As big an assortment as your space and budget will allow. Ideally, you'll replace spices annually, more or less; even whole spices don't keep forever. But they're inexpensive, especially if you buy them in ethnic markets rather than supermarkets, or from a specialist like Penzey's (www.penzeys.com). Chiles are a special case (though you can usually buy them where you buy spices). Stock some dried hot ones (usually red) and also some mild ones, like pasillas. Chipotles are good to have around, too, for their unique smoky-hot flavor.

Vinegar. I think sherry vinegar is the most versatile and best for the money; balsamic, of course, is more popular, but sweeter. If you can find well-made red and white wine vinegar, those are good too, as is rice vinegar, which has the lowest acidity of any vinegar and important in Asian-style dishes.

Soy sauce. Look for brands that contain no more than soy, wheat, salt, water, and bacteria.

Dried fruit and nuts. For snacking and for cooking. The fat in nuts can go rancid, so don't buy more than you're going to use in a month or so (or store in the freezer). Remember sesame and sunflower seeds too, and nut butters, including tahini, which makes an excellent sauce when mixed with a little water or lemon juice.

Meat, dairy, and cheese. The whole argument here, of course, is that you don't need meat or dairy food, but most people are going to want some of it around (I do, anyway), and the following four things all keep well, are very useful, and add great flavor whenever you use them. They're also obvious, but hey:

- **Bacon.** Keep a hunk in the freezer or fridge and use it for seasoning. An ounce goes a long way.
- **Parmesan cheese.** Lasts forever; grated over almost any salad or pasta dish, is just a killer.
- **Butter.** What can I say? As an occasional alternative to olive oil in cooking or flavoring, a real pleasure.
- **Eggs.** Possibly the most useful of all animal products.

Canned tomatoes. Couldn't be simpler. Plum tomatoes tend to be sweetest and best-tasting. Pre-chopped tomatoes make life a little easier. Avoid those with additives.

Sweeteners. Sugar is unavoidable and of course fine in moderation. But maple syrup and honey are in a way far more useful, since they deliver more flavor along with sweetness.

Baking soda, baking powder, instant yeast. Especially if you're into baking.

The Advanced Food Matters Pantry

You don't need everything here every day, or perhaps not even every week, but most of it keeps well and many items, once exotic, have become part of contemporary eating. Many of them provide instant flavor.

Dried mushrooms. Especially porcini (cèpes) and shiitakes.

Capers. Packed in vinegar or salt.

Miso. Truly one of the world's great ingredients; instantly adds depth to soups and stews, and you can use it to make quick sauces and dressings simply by adding water.

Anchovies. Packed in olive oil: best bought in resealable glass jars rather than cans.

Sesame oil. The roasted kind, sold in all Asian markets. No better way to finish a stir-fry.

Sea Greens (seaweed). Especially hiziki, konbu (kelp), and nori (laver).

Coconut milk. Easily made fresh, but far more convenient when canned. Light is an acceptable option in almost every case.

Silken Tofu. The kind in the box keeps for months in the pantry. It's a handy substitute for sour cream or other thickeners, and it is also nice slipped into a bowl of soup or tossed with Asian noodles and vegetables.

1
WEEK 49
7
8
9
14
WEEK 51
22

Meal Plans: A Sample Month

As I've said repeatedly, the Food Matters style of eating isn't dogmatic. Beyond "Eat fewer animal products and more plants" there are really no rules, so providing meal plans feels a bit counterintuitive.

Nevertheless, there are good reasons for giving you some guidelines. For one thing, people I talk to are interested in hearing about what I eat in a normal day; once they learn that many dinners still include sausage, steak, pasta, or roast chicken, they're less intimidated. I've also discovered that many people really want to see what a typical day or week looks like before they consider a change in their dietary behavior, which makes sense. There's also an expectation that a book like this will have meal plans; that's fair enough.

So that's what the next few pages will give you: a sense of the Food Matters diet in practice. If you're someone who likes more concrete directives, these can help you get started. But even if you might resent being told what to eat when, they can still serve as an inspiration for integrating the basic concepts and recipes into your daily life on your own terms.

I've included suggested snacks—they sound formal—but remember that throughout the day, every day, you can (and probably should) be eating fruit, nuts, raw and cooked vegetables, popcorn, salads, vegetable soups, and the like.

And keep in mind that these are all *suggestions,* no more: Use what you can, but make it simple and seasonal. If peaches aren't in season, make a banana smoothie instead. If the eggplant looks like it's seen better days, there's nothing wrong with zucchini and chicken Parmesan; or make paella with scallops instead of shrimp. Remember, this is all about what you *can* eat, not what you can't.

Week 1

Monday

BREAKFAST
Blueberry smoothie (page 163)

Whole grain toast

LUNCH
Lunchtime Caesar Salad with shrimp (page 184)

SNACK
Brown-Bag Popcorn (page 235) with Sesame Shake (page 145)

DINNER
Stuffed Chicken Breasts with Pan-Grilled Corn (page 268)

Steamed asparagus (page 133) with Olive Oil Drizzle (page 158)

Wild rice (page 136)

DESSERT
Lemon sorbet (page 298)

Tuesday

BREAKFAST
Breakfast Burritos (page 176) with beans (page 139) and avocado

LUNCH
Asian-Style Noodles with Mushrooms (page 213)

SNACK
Warm Nuts and Fruit (page 218)

DINNER
Roasted Vegetables with halibut or salmon steaks (page 241)

Couscous (page 137)

DESSERT
Fresh orange wedges

Wednesday

BREAKFAST
Breakfast Couscous (page 166) with Roasted Vegetables (page 134)

LUNCH
Salade Niçoise with Mustard Vinaigrette (page 186)

Hybrid Quick Bread (page 154)

SNACK
Banana smoothie (page 163)

DINNER
Bean and Vegetable Chili (page 260)

Brown rice (page 136)

Fresh tomato salsa (page 146)

DESSERT
Nutty Oatmeal Cookies (page 290)

Thursday

BREAKFAST
Fruit salad (page 162)

Bulgur with maple syrup and dried fruit (pages 137, 169)

LUNCH
Japanese Mixed Rice (page 253)

SNACK
Baked Pita Triangles (page 234) with spinach spread (page 222)

DINNER
Eggplant and Chicken Parmesan (page 270)

Farro with Olive Oil Drizzle (pages 136, 158)

DESSERT
Fresh or frozen berries

Friday

BREAKFAST
Granola with yogurt and mixed berries (page 168)

LUNCH
Thai Beef Salad (page 188)

SNACK
Brown-Bag Popcorn (page 235) **with Hot Curry Powder** (page 144)

DINNER
Easy Whole Grain Pizza with caramelized onions and mushrooms (page 225)

Nicely Dressed Salad Greens with chickpeas and cucumber (page 135)

DESSERT
No-Bake Apricot Fruit Tarts (page 286)

Saturday

BREAKFAST
Better Poached Eggs with asparagus (page 171)

LUNCH
Hummus with Pita and Greens (page 196)

SNACK
Whole grain toast with carrot spread (page 222)

DINNER
Savory Vegetable and Grain Torta (page 280)

Mesclun salad (page 135)

DESSERT
Chocolate Semolina Pudding with Raspberry Puree (page 296)

Sunday

BRUNCH
Fruit salad (page 162)

Whole Grain Pancakes (page 172)

SNACK
Brown-Bag Popcorn (page 235) **with Med Mix** (page 172)

DINNER
Modern Bouillabaisse (page 275)

Arugula salad with toasted walnuts (page 135)

Toasted whole grain baguette slices

DESSERT
Peach sorbet (page 298)

Fruit and Cereal Bites (page 232)

Week 2

Monday

BREAKFAST
Breakfast Burritos (page 176) **with black beans** (page 139) **and peach salsa** (page 146)

LUNCH
Spinach and Sweet Potato Salad with Warm Bacon Dressing (page 194)

SNACK
Fruit and Cereal Bites (page 232)

DINNER
Herb-stuffed eggplant (page 272)

Shaved fennel and apple salad (page 135)

Couscous (page 137)

DESSERT
Nutty Oatmeal Cookies (page 290)

Tuesday

BREAKFAST
Banana-peach smoothie (page 163)

Couscous with raisins and dates (page 166)

LUNCH
Fast Mixed Vegetable Soup (page 200)

Whole grain crackers

SNACK
Beet chips (page 226)

DINNER
Meat loaf with bulgur and ground beef (page 278)

Sautéed carrots and parsnips (page 133)

Pot of chickpeas or fava beans (page 139)

DESSERT
Roasted apples and figs (page 292)

Wednesday

BREAKFAST
Oatmeal (page 164) **with roasted figs and chopped pecans** (page 292)

LUNCH
Vegetable Fried Rice (page 208)

SNACK
Crisp Nori Ribbons (page 227)

DINNER
Baked Ziti (page 252)

Mesclun salad (page 135) **with Roasted Red Peppers** (page 152)

DESSERT
Mango smoothie (page 163)

Thursday

BREAKFAST
Granola with yogurt and bananas (page 168)

LUNCH
Stir-Fried Greens with Cashews (page 212)

SNACK
Baked Pita Triangles (page 234) **with carrot-parsnip spread** (page 222)

DINNER
Curried Lentil Stew with Chicken and Potatoes (page 204)

Winter greens with fresh pears (page 135)

DESSERT
Chocolate-cherry sorbet (page 298)

Friday

BREAKFAST
Whole grain toast with spinach spread (page 222)

LUNCH
Ratatouille, in a whole grain wrap (page 206)

SNACK
Brown-Bag Popcorn (page 235) **with Fragrant Curry Powder** (page 144)

DINNER
Braised root vegetables with pork (page 243)

Steamed spinach (page 132)

Quinoa (page 136)

DESSERT
No-Bake Fig Tarts (page 286)

Saturday

BREAKFAST
More-Vegetable-Than-Egg Frittata (page 170)

Hybrid Quick Bread (page 154)

LUNCH
Steamed Asian vegetables (page 132) **with Sesame Oil Drizzle** (page 158)

Brown rice (page 136)

SNACK
Baked tortilla chips (page 234) **with tomatillo salsa** (page 146)

DINNER
Grilled vegetable and meat shish kebabs (page 246)

Tabbouleh My Way (page 190)

DESSERT
Chocolate Fondue with Fresh Fruit (page 288)

Sunday

BRUNCH
Fruit salad (page 162)

Breakfast Bread Pudding (page 174)

SNACK
Crudités You Actually Want to Eat (page 220)

DINNER
Cassoulet with Lots of Vegetables (page 262)

Salad greens with dried apricots and almonds (page 135)

Toasted whole grain baguette slices with Fast Roasted Garlic (page 159)

DESSERT
Brown Rice Pudding with coconut (page 294)

Week 3

Monday

BREAKFAST
Oatmeal with maple syrup and dried fruit (page 164)

LUNCH
Creamy Carrot Soup (page 202)

Easy Socca or Farinata (page 225)

SNACK
Mixed nuts

DINNER
Stir-Fried Vegetables with shrimp and scallops (page 238)

Quinoa (page 136)

DESSERT
Brown Rice Pudding with mango (page 294)

Tuesday

BREAKFAST
Quinoa with honey and nuts (pages 136, 169)

LUNCH
Whole Grain Bread Salad (page 192)

SNACK
Mango smoothie (page 163)

DINNER
Bulgur Pilaf with Vermicelli, and ground lamb (page 258)

Steamed cauliflower (page 132) **with Olive Oil Drizzle** (page 158)

Sautéed spinach (page 133)

DESSERT
Apple-cranberry crisp (page 292)

Wednesday

BREAKFAST
Swiss-Style Muesli with fresh fruit (page 169)

LUNCH
Chopped Cabbage Salad, Asian Style (page 181)

SNACK
Crisp Nori Ribbons (page 227)

DINNER
Pan-cooked grated vegetables with crunchy salmon (page 248)

Brown rice with shallots (page 136)

DESSERT
Fruit and Cereal Bites (page 232)

Thursday

BREAKFAST
Breakfast Burritos (page 176) **with brown rice** (page 136) **and lettuce**

LUNCH
Zucchini pancakes (page 230) **on Nicely Dressed Salad Greens** (page 135)

SNACK
Fruit and Cereal Bites (page 232)

DINNER
Orchiette with Broccoli Rabe, My Style (page 251)

Cherry tomatoes salad with fresh basil (page 135)

DESSERT
Fresh melon wedges or broiled grapefruit

Friday

BREAKFAST
Swiss-Style Muesli with dried fruit and nuts (page 169)

LUNCH
Layered Peach Salad (page 183)

SNACK
Pineapple-coconut smoothie (page 163)

DINNER
Asian-Style Noodles with Mushrooms (page 215)

Napa cabbage and snow pea salad (pages 180–81)

DESSERT
Coconut and Nut Chews (page 289)

Saturday

BREAKFAST
Strawberry-banana smoothie (page 163)

Whole grain toast

LUNCH
Stir-fried spinach with tofu and garlic (page 211)

SNACK
Baked Tortilla Chips (page 234) with papaya salsa (page 146)

DINNER
Roasted winter vegetables with lamb chops (page 241)

Pot of big lima beans (page 139)

DESSERT
Melon sorbet (page 298)

Coconut and Nut Chews (page 289)

Sunday

BRUNCH
Whole Grain Pancakes with fresh fruit (page 172)

SNACK
Big beans on toothpicks (page 228)

DINNER
Chicken Not Pie (page 266)

Romaine salad (page 135) with baked croutons (page 234)

DESSERT
Chocolate Semolina Pudding with Raspberry Puree (page 296)

Week 4

Monday

BREAKFAST
Breakfast Couscous, with almonds and dried cherries (page 166)

LUNCH
Stir-fried beans with broccoli (page 199)

SNACK
Brown-Bag Popcorn (page 235) with Five-Spice Powder (page 152)

DINNER
Cornmeal pizza with sausage and mushrooms (page 225)

Mesclun salad with cucumber and fennel (page 135)

DESSERT
Mango smoothie (page 163)

Tuesday

BREAKFAST
Breakfast Burritos (page 177) with mango salsa (page 146)

LUNCH
Chunky and Creamy Carrot Soup (page 203)

Easy Whole Grain Flatbread (page 224)

SNACK
Blueberry smoothie (page 163)

DINNER
Paella with shrimp and chorizo (page 256)

Salad greens with radishes (page 135)

DESSERT
Nutty Oatmeal Cookies (page 290)

Wednesday

BREAKFAST
Fruit salad (page 162)

Swiss-Style Muesli with yogurt (page 169)

LUNCH
Lunchtime Caesar Salad with chicken (page 184)

SNACK
Sweet potato chips (page 226)

DINNER
Lentil and lamb burger (page 278)

Not Your Usual Ratatouille (page 206)

Pot of chickpeas (page 139)

DESSERT
Blueberry sorbet (page 298)

Thursday

BREAKFAST
Fruit salad (page 162)

Whole grain toast

LUNCH
Hummus with Pita and Greens (page 196)

SNACK
Peach smoothie (page 163)

DINNER
Spaghetti with Tomato Sauce (page 147)

Baby artichoke salad with Parmesan (page 135)

DESSERT
No-bake raspberry tarts (page 286)

Friday

BREAKFAST
Fruit salad (page 162)

Swiss-Style Muesli (page 169)

LUNCH
Quick Vegetable Fried Grain (page 209)

SNACK
Baked Pita Triangles (page 234) **with Hummus** (page 196)

DINNER
Herb-stuffed acorn squash (page 272)

Baby spinach with hazelnuts and dried cranberries (page 135)

White beans (page 139) **with olive oil and Med Mix** (page 144)

DESSERT
Plum sorbet (page 298)

Coconut and Nut Chews (page 289)

Saturday

BREAKFAST
More-Vegetable-Than-Egg Frittata (page 170)

Hybrid Quick Bread (page 154)

LUNCH
Chile Mixed Rice (page 254)

SNACK
Baked Pita Triangles (page 234) **with white bean spread** (page 222)

DINNER
Savory Vegetable and Grain Torta (page 280)

Endive salad with red and yellow peppers (page 135)

DESSERT
Brown Rice Pudding with dried cherries (page 294)

Sunday

BRUNCH
Celery root and apple pancakes (page 230)

Hybrid Quick Bread (page 154)

SNACK
Crudités with *Bagna Cauda* (page 221)

DINNER
Cassoulet with Lots of Vegetables (page 262)

Arugula salad with shaved Parmesan cheese (page 135)

DESSERT
Apricot and cherry crisp (page 292)

The Basics

There are a number of basic recipes that no cook can really do without, but in the world of sane eating, a few of these are absolutely essential: You must be able to make a pot of beans, a bowl of cooked grains, a simple steamed vegetable, a salad. This chapter will show you how. If you're a beginning cook, these are true basics; if you're a veteran, there are some twists that might be new to you.

This chapter also includes salsas, drizzles, and spice blends, with enough variations so that you can change the flavor profile of just about anything you cook without much effort. Once you have your pantry set up, it's easy to create basic flavor combinations that can make even the simplest food—lettuce leaves, steamed broccoli, boiled grains, cut up (or grated) raw veggies—taste not only delicious but novel. Most of these can be made once and then used whenever you need them, weeks or even months later.

Beyond the staples are some building blocks that you'll use over and over: tomato sauce, in a few different guises, none of which take more than a half hour or so; the new classic, roasted garlic, and the old standby, roasted peppers; stocks, long considered the backbone of fine cuisine and still useful in many contemporary dishes; a nearly whole-grain quick bread (45 minutes, start to finish, no kidding), and a revolutionary whole grain yeasted loaf that takes some time to ferment but almost no work and fits in perfectly with the Food Matters style of eating.

Boiled or Steamed Vegetables, As You Like 'Em

Makes: **4 servings** Time: **10 to 30 minutes**

Preparing vegetables shouldn't be a big deal, and in the Food Matters kitchen, all stages of doneness have their place. This all-purpose recipe allows you to control their texture, no matter what type of vegetable you cook or how tender you want it.

Barely cooked vegetables are perfect for fresh-tasting salads, or if you want to reheat them in stir-fries or on the grill, where they'll soften up a bit more. Crisp-tender vegetables can be brought directly to the table or plunged into a bowl of ice water to capture their perfect texture; they can be reheated, too. Soft-cooked vegetables can provide instant pleasure, especially when pureed or mashed (I love them on a thick piece of whole wheat toast).

Two Technical Notes: To rig a steamer, fit a steaming basket in a large pot with a tight-fitting lid and add water so that the basket sits above it. If you don't have a basket, use two ovenproof plates: put the first one facedown in the pot and the second faceup. Fill the pot with enough water to submerge the plate on the bottom; use the top plate to hold the food.

If you're cooking thick-stemmed greens like bok choy, kale, collards, or even broccoli, consider separating the stems from the leaves (or florets in the case of broccoli), and toss the stems into the pot a couple of minutes before the greens so everything reaches about the same degree of doneness.

Salt

About 2 pounds of virtually any vegetable (including greens), peeled, stemmed, seeded, and/or chopped as needed

Freshly squeezed lemon juice or olive oil, as you like

Freshly ground black pepper

Chopped fresh herbs or ground seasonings (optional)

1 Bring a pot of water to a boil and salt it, or rig a steamer as described in the note. If you want to "shock" the vegetables to capture doneness at a precise moment, fill a large bowl (or a clean sink) with ice water. Try bending or breaking whatever it is you're planning to cook: the more pliable the pieces are, the more quickly they will become tender.

2 When the water boils, add the vegetables to the pot or steamer. Check tender greens in less than a minute; root vegetables (which are usually best completely tender) will take 10 minutes or more. Everything else is somewhere in between. Every so often while the vegetables are cooking, use tongs to grab a piece out and test it. (With experience, you'll do this less frequently.) Remember that unless you shock them in ice water, the vegetables will continue to cook, and become more tender even when they're off the heat.

3 When the vegetables are cooked as you like them, drain and serve, drizzled with lemon juice, oil, butter, more salt and pepper, or whatever. Or plunge the drained vegetables immediately into the ice, drain again, and set aside to use later.

Sautéed Vegetables: You can boil or steam the vegetables first, then sauté them; or cook them, starting raw, directly in the oil. Put a film of olive oil in the bottom of a large skillet and turn the heat to medium-high. When it's hot, add the vegetables, sprinkle with salt and pepper, and cook, stirring occasionally and checking for doneness as described in Step 1. (The only difference is that you'll be fishing test pieces out of a skillet, not out of a pot of boiling water.) Precooked vegetables are ready as soon as they're rewarmed; tender greens will take 5 to 10 minutes; and cubed root vegetables up to half an hour. When they're ready, taste and adjust seasoning (add fresh herbs or spices and lemon juice if you want), stir, and serve hot or at room temperature.

Roasted Vegetables: Heat the oven to 425°F. Put the vegetables, alone or in combination (even greens work here), in a large roasting pan or on a rimmed baking sheet and toss them with at least 3 tablespoons of olive oil. Sprinkle with salt and pepper, and start roasting. Check tender vegetables in 10 minutes, sturdier ones in 15. Whenever you check, turn or stir as necessary to promote even cooking. Total time will be between 15 and 45 minutes, depending on the size and type of vegetable. When the vegetables are ready, taste and adjust seasoning (add fresh herbs or spices and lemon juice if you want), toss, and serve hot or at room temperature.

Grilled or Broiled Vegetables: Even sturdy greens like radicchio and romaine work here; just quarter or halve them, leaving the root ends intact. Heat a broiler or grill and move the rack about 4 inches from the heat source. Put the vegetables in a large bowl, toss them with at least 2 tablespoons of olive oil, and sprinkle with salt and pepper. When the fire is hot, put them on the grill or in a pan under the broiler. Start checking tender vegetables in a minute or two; sturdier vegetables in 10. Turn and move them around as necessary to promote even cooking. Total time will be between 5 and 20 minutes, depending on the size and type of vegetable. When the vegetables are ready, sprinkle with more salt and pepper, a squeeze of lemon, herbs, or spices.

Nicely Dressed Salad Greens (or Anything Else)

Makes: **4 servings** Time: **10 minutes**

All you need to make an excellent dressing are a flavorful oil, a little vinegar or lemon juice, some salt and pepper, and a bowl. Don't think of this as only a treatment for salad, it's also for sliced raw or grilled vegetables, plain cooked grains, beans, greens, or other veggies—or all of the above. The technique remains the same.

I usually start with extra-virgin olive oil and sherry vinegar, but why not try grape seed oil with a few drops of sesame oil and rice vinegar for a salad that goes with Asian dishes? Add high-flavor ingredients (like spices or chopped herbs, onions, garlic, ginger, or olives) along with the oil and vinegar. Add tomatoes, bits of meat or fish, nuts, and so on, before you toss, or serve them on top of the dressed greens. And if you're dressing cooked food, dress it while still warm so it absorbs flavor as it cools.

- 8 cups assorted greens or other raw or cooked plant foods
- ⅓ cup olive oil, more or less
- 2 tablespoons sherry vinegar or balsamic vinegar or freshly squeezed lemon juice, more or less
- Salt and freshly ground black pepper

Put the greens in a bowl and drizzle them with most of the oil and vinegar. Sprinkle with salt and pepper. Toss a few times and taste. Adjust the seasoning, adding more oil or vinegar if you like; toss again, and serve immediately.

Whole Grains without Measuring

Makes: **8 servings (6 to 8 cups)** Time: **10 minutes to more than 1 hour, depending on the grain**

One sure way to eat more whole grains is to always have some handy, and that's almost as easily said as done: they bubble along without fuss, can be made ahead, and keep in the fridge for about a week. Once you get used to making whole grains, you'll probably stop measuring. That's fine: the amount of water grains will absorb varies anyway, depending on their age and how they were stored.

My technique works for just about anything, including rice (the exceptions are noted in the variations). You won't normally eat this whole batch, but again, grains keep and reheat perfectly. If you want to cook smaller quantities more frequently, just cut the amount of grains in half, more or less.

For more about what to do with the grains once they're cooked, see the list on page 138.

2 cups brown rice (any size), quinoa, barley (any type), oat groats, buckwheat groats, steel-cut oats, millet, cracked wheat, hominy, whole rye, farro, kamut, or wild rice; or 1½ cups wheat berries

Salt

Olive oil, or other vegetable oil (optional)

1 Rinse the grain in a strainer, and put it in a large pot with a tight-fitting lid along with a big pinch of salt. Add enough water to cover by about an inch (no more); if you want your grains on the dry side, cover with closer to ½ inch of water. Use 3 cups water for pearled barley, which absorbs a more precise amount of water. Bring to a boil, then adjust the heat so the mixture bubbles gently.

2 Cook, stirring once in a while, until the grain is tender. This will take as little as 7 or 8 minutes for steel-cut oats, about 40

minutes for brown rice, and as long as 1 hour or more for some specialty rices, unpearled or hulled barley, wheat berries, and other unhulled grains. Add boiling water as necessary to keep the grains just submerged, but—especially as the grain swells and begins to get tender—keep just enough water in the pot to prevent the grain from drying out and sticking.

3 Every now and then, test a grain. Grains are done when just barely tender (they should always have some chew). Be careful not to overcook unless you want them on the mushy side. If the water is all absorbed at this point (one sure sign is that little holes have formed in the top), then cover and remove from heat. If some water still remains, drain the grains (reserving the water for soup if you like), and immediately return the grains to the pot, cover, and remove from the heat. Either way, undisturbed, they'll stay warm for about 20 minutes.

4 Toss the grain with the oil if you like, and refrigerate or freeze until ready to use. If you're serving it right away, see What to Do with Cooked Grains (next page).

Couscous: Put 2 cups of whole wheat couscous in a medium pot with a tight-fitting lid and add 3 cups of water and a pinch of salt. Bring the water to a boil, then cover and remove from the heat. Let steep for at least 10 minutes (5 minutes if using white couscous), or up to 20. Fluff with a fork and serve as described in the main recipe.

Bulgur: Put 2 cups of bulgur (any grind) in a large bowl with a pinch of salt. Pour 5 cups boiling water over all. Stir once and let sit. Fine bulgur will be tender in 10 to 15 minutes, medium in 15 to 20 minutes, and coarse in 20 to 25. If any water remains when the bulgur is done, put the bulgur in a fine strainer and press down on it, or squeeze it in a cloth. Fluff with a fork and serve as described in the main recipe.

What to Do with Cooked Grains

In Step 4, use a fork to toss any of these ingredients—alone or in combination—in with the grains and the oil:

Cooked vegetables (ideally crisp-tender), like peas, chopped greens, broccoli or cauliflower florets, or chopped carrots or other root vegetables.

A couple of spoonfuls of sauce, like any flavored olive oil (see Olive Oil Drizzle, page 158) or any tomato sauce (see All-Purpose Tomato Sauce, page 147) or a bottled condiment like soy sauce or hot sauce.

A sprinkling of chopped fresh herbs, like chives, parsley, cilantro, or mint; or a bit of rosemary, oregano, or thyme.

Any cooked beans (as many as you like).

Dried fruit, like raisins, cranberries, cherries, or chopped dates or apricots, with or without chopped nuts or seeds.

Cooked chopped sausage, bacon, ham, or any cooked meat or fish.

Pot of Beans

Makes: **6 to 8 servings** Time: **1 to 2 hours to soak plus 30 minutes to 2 hours to cook, depending on the bean, largely unattended**

I'm on a mission to make sure every fridge or freezer in America is stocked with a container of home-cooked beans, and this recipe is my ammunition—a simple process that requires no advance planning and very little attention, yet provides the backbone for several delicious meals.

Here are foolproof beans any way you like them: skins intact for salads and stir-fries, or soupy for spooning over rice. If you have time to soak the beans without boiling them, put them in a bowl with tap water to cover and set them aside (no longer than 12 hours, or they'll cook up mushy). You can also skip soaking altogether and cook the beans straight through; it won't take much longer.

Some people believe a pot of beans has no flavor without some meat. I disagree, but meat certainly adds richness; you might, however, be surprised at how little meat it takes to do the trick. See the sidebar on pages 141–42 for some flavoring and serving suggestions.

1 pound dried beans (any kind but lentils, split peas, or peeled and split beans), washed and picked over

Salt and freshly ground black pepper

1 Put the beans in a large pot with a tightly fitting lid and cover with cold water by a couple of inches. Bring the pot to a boil and let it boil, uncovered, for about 2 minutes. Cover the pot and turn the heat off. Let the beans soak for at least 1 hour or up to 2 hours.

2 Taste a bean. If it's at all tender (it won't be ready), add a large pinch of salt and several grinds of black pepper. Make sure the beans are covered with about an inch of water; add a little more if necessary. If the beans are still hard, don't add salt yet, and cover with about 2 inches of water.

3 Bring the pot to a boil, then adjust the heat so that the beans bubble gently. Partially cover and cook, stirring every now and then, checking the beans for doneness every 10 or 15 minutes, and adding more water if necessary, a little at a time. Small beans will take as little as 30 minutes more; older, large beans can take up to an hour or more. If you haven't added salt and pepper yet, add them when the beans are just turning tender. Stop cooking when the beans are done the way you like them, and taste and adjust the seasoning.

4 Here you have a few options. Drain the beans (reserving the liquid separately) to use them as an ingredient in salads or other dishes where they need to be dry; or finish them with one of the ideas from the list below. Or store the beans as is and use with or without their liquid as needed. They'll keep in the fridge for days, and in the freezer for months.

Pot of Lentils or Split Peas: No need to soak, since they cook fast—usually in less than 30 minutes. Put them in a large pot with a tightly fitting lid and cover with cold water by a couple of inches. Bring the pot to a boil, then reduce the heat so that they bubble gently. Partially cover and cook, stirring infrequently and checking for doneness every 10 or 15 minutes; add a little more water if necessary. When they start to get tender, add a large pinch of salt and several grinds of black pepper; stop cooking when they're done the way you like them, taste and adjust the seasoning, and use immediately or store.

Pot of Fresh (or Frozen) Shell Beans: For limas, favas, edamame (in or out of the pod), and the like; cook like vegetables. Bring a pot of water to boil and salt it. Add the beans and cook until just tender (if the beans are in their shells, test one every now and then). This can be as quick as a few minutes, and rarely does it take longer than 10 minutes. Drain, reserving the liquid if you like, and serve or refrigerate for later.

Adding Flavor to a Pot of Beans

Here are some goodies to add (alone or in combination) when you start cooking the beans.

Herbs or spices: a bay leaf, a couple of cloves, some peppercorns, thyme sprigs, parsley leaves and/or stems, chili powder (to make your own, see page 143), or other herbs and spices.

Aromatics: Chopped onion, carrot, celery, and/or garlic. If you like, sauté them in a little olive oil until soft and fragrant.

Quick Vegetable Stock (page 150), in place of all or part of the water.

Other liquids: A cup or so of beer, wine, coffee, tea, or juice.

Smoked meat: Ham hock, pork chop, beef bone, or sausage, fished out after cooking, the meat chopped and stirred back into the beans.

Add any of these ingredients after you cook and drain the beans; the quantities listed work for about 3 cups of cooked beans, or 4 servings. You might reheat gently to blend flavors, adding the reserved cooking liquid if needed to keep the beans moist.

2 tablespoons olive or dark sesame oil

½ cup chopped fresh parsley, cilantro, mint, or any basil leaves

2 tablespoons chopped fresh rosemary, tarragon, oregano, epazote, thyme, marjoram, or sage leaves

Chopped scallions, garlic, ginger, or lemongrass

1 cup any cooked sauce (like All-Purpose Tomato Sauce, page 147) or raw sauce (like Salsa, Any Style, page 146)

1 tablespoon or so of anything listed in Six Seasoning Blends You Can't Live Without (page 143)

Soy, Worcestershire, or hot sauce to taste

A couple of tablespoons miso thinned with hot bean-cooking liquid

Chopped leafy greens, like spinach, kale, or collards

Peeled, seeded, and chopped tomato

1 or 2 slices of diced bacon (or pancetta), or a fresh crumbled sausage, cooked until crisp (along with some of the fat if you like)

Six Seasoning Blends You Can't Live Without

Makes: **About 1/4 cup of each** Time: **No more than 10 minutes**

You can make any simple food taste great if you have your own spice mix lying around; a little sprinkle of this on grains, a tablespoon or two in beans—it's the contemporary equivalent of ketchup on a burger. Start with preground spices, or, for even better-tasting results, toast and grind your own from whole ingredients.

Stored in opaque containers in a cool place, these mixtures will last for several weeks. Use along with oil and aromatic vegetables like onions, garlic, or ginger when you're cooking; or just rub into food before grilling, broiling, or roasting; or sprinkle on anything just before serving, as you would salt and pepper.

If you're using whole spices, put them in a dry skillet over medium heat. Toast, shaking the pan occasionally, until the mixture is fragrant, 3 to 5 minutes, lowering the heat if necessary so as not to scorch them. Add any ground spices during the last minute of cooking. Then grind everything together in a spice or coffee grinder until powdery.

Chili Powder

2 tablespoons ground ancho, New Mexico, or other mild dried chile

1/2 teaspoon cayenne, or to taste

1/2 teaspoon black peppercorns

2 teaspoons cumin seeds

2 teaspoons coriander seeds

1 tablespoon dried oregano (preferably Mexican)

Med Mix (toast only a minute or so)

2 bay leaves

1 tablespoon dried rosemary leaves

1 tablespoon dried sage leaves

1 tablespoon dried thyme leaves

1 tablespoon dried parsley or lavender leaves

Hot Curry Powder

2 small dried Thai or other hot chiles

1 tablespoon black peppercorns

1 tablespoon coriander seeds

1 teaspoon cumin seeds

1 teaspoon fennel seeds

1 teaspoon ground fenugreek

1 tablespoon ground turmeric

1 tablespoon ground ginger

Cayenne or red pepper flakes, as needed

Fragrant Curry Powder

¼ teaspoon nutmeg pieces

Seeds from 5 white cardamom pods

3 cloves

One 3-inch cinnamon stick

1 teaspoon black peppercorns

2 tablespoons cumin seeds

¼ cup coriander seeds

2 bay leaves

2 dried curry leaves (optional)

1 teaspoon ground fenugreek

Five-Spice Powder

1 tablespoon Sichuan peppercorns or black peppercorns

6 star anise

1½ teaspoons whole cloves

One 3-inch stick cinnamon

2 tablespoons fennel seeds

Sesame Shake

¼ cup sesame seeds

1 tablespoon dried thyme leaves or 2 tablespoons dried seaweed, like dulse or arame

Cayenne, to taste

Salsa, Any Style

Makes: **About 4 cups** Time: **20 minutes**

The formula for salsa is consistent, whether you start with the usual tomatoes, or the less expected radishes, cherries, or something else. Replace some or all of the tomatoes with fruit—peaches, nectarines, plums, pineapple, papaya, mango, tart green apples, grapes, citrus, berries, and on and on. Crisp vegetables (like raw fresh corn, jicama, or bell peppers) add nice texture. And you can change the flavor profile just as easily: Use basil and balsamic vinegar instead of cilantro and citrus juice; or go off in the direction of rice vinegar and curry. The combinations are infinite.

You can also fool around with the texture by pureeing half or all of the salsa for smooth or smooth-chunky versions. In short, there are almost no rules here.

4 large ripe fresh tomatoes, cored and chopped (about 3 cups)

1 large white onion or 5 or 6 scallions, chopped

2 teaspoons minced garlic, or to taste (optional if you're using all fruit)

Minced fresh chile (like jalapeño, Thai, or less of habanero) or hot red pepper flakes or cayenne to taste

1 cup chopped fresh cilantro or parsley leaves

4 tablespoons freshly squeezed lime juice or 2 tablespoons sherry or wine vinegar

Salt and freshly ground pepper

Combine everything but the salt and pepper in a bowl. Sprinkle with salt and pepper, then taste and adjust the seasoning. If you have time, let the flavors mingle for 15 minutes or more before serving.

All-Purpose Tomato Sauce

Makes: **6 to 8 servings (about 1 quart)** Time: **30 minutes**

A batch of tomato sauce is fast and easy. If you don't finish it in a few days you can freeze what's left; but I try to keep some in the fridge at all times, since it reheats well and is good on everything from steamed vegetables to simply cooked fish or chicken, and of course pasta or rice. You can also turn the tables and make the sauce the base for braised string beans, tofu, celery, or any combination: Parboil the vegetables if necessary, then finish cooking them in the sauce.

Note: You can buy prechopped tomatoes, or just chop them in the cans by swirling a knife through them. See the first variation for fresh sauce.

¼ cup olive oil

1 large onion or 2 medium onions, chopped

About 4 pounds canned whole tomatoes (two 28- or 35-ounce cans), chopped, liquid reserved

Salt and freshly ground black pepper

½ cup chopped fresh parsley or basil leaves (optional)

1 Put the olive oil in a pot over medium heat. When the oil is hot, add the onions, sprinkle with salt and pepper, and cook, stirring occasionally, until soft, about 3 minutes. Then add the tomatoes.

2 Cook, stirring occasionally, until the tomatoes break down and the mixture comes together and thickens a bit, 10 to 15 minutes. For a thinner sauce, add some or all of the reserved liquid and cook for another 5 to 10 minutes; if you want a thick sauce, save it for another use. Taste, adjust the seasonings, stir in the herbs, and keep warm. (Or let cool, cover, and refrigerate for up to several days; reheat gently before serving.)

Five Flavor Boosters for Basic Tomato Sauce

Add any of these to the onions, right before stirring in the tomatoes:

- Chopped black or green olives, and/or capers, and/or anchovies
- Red pepper flakes or a whole dried chile (fish it out later)
- A couple of bay leaves (fish them out later)
- 1 ounce (or more) reconstituted dried porcini mushrooms (page 213)
- The rind from a wedge of Parmesan cheese

Fresh Tomato Sauce: Instead of canned tomatoes, use chopped fresh (peeled and seeded if you like, but if not, so be it).

Homemade Cooked Salsa: For either the main recipe or the fresh variation above. In Step 1, after the onions cook for about 2 minutes, add 2 or more chopped fresh jalapeño, serrano, or other fresh hot chiles (with the seeds if you like; they're hot), along with 2 tablespoons chopped garlic. Instead of parsley or basil, finish with chopped cilantro.

Garlicky Tomato Sauce: Omit the onions. Chop 4 to 10 (or more) raw garlic cloves, or use 1 or 2 heads Fast Roasted Garlic (page 159; squeezed from the skins) and add them to the hot oil in Step 1. Reduce the cooking time to just a minute or so. Proceed with the recipe.

All-Purpose Tomato Sauce Spiked with Sausage or Meat: In Step 1, before stirring in the onions, brown a chopped Italian sausage, or ¼ pound ground meat; add the onions when the meat is just starting to loose its pink, and proceed with the recipe.

Peeling Fruit (Including Tomatoes)

To remove the skins from any fruit, get a pot of water boiling and set up a bowl of ice water near the stove. Immerse the fruit, one or a few pieces at a time, in boiling water for about 30 seconds, or until the skin loosens, then plunge into ice water to stop the cooking. The skin should peel off easily with a paring knife.

Quick Vegetable Stock

Makes: **More than 2 quarts** Time: **20 to 40 minutes**

Homemade stock is not only infinitely better than the packaged stuff but also easy to make. You can produce large quantities and freeze enough to have it on hand for soup, rice, beans—or anytime when you want more flavor than water.

Making stock is more art than science, so don't worry if you don't have all the ingredients here; use what you have, and substitute at will. Off-limits are strong-flavored greens (cabbage included—unless you want your stock to taste like cabbage), bell peppers, and eggplant. Don't even bother to peel the vegetables, just wash the skins well.

For a more robust, earthy flavor, toss in a handful of dried porcini (or sautéed fresh mushrooms). Dried or canned tomatoes add color and flavor. Or, try a couple of bay leaves and a few sprigs of fresh thyme for a more fragrant variation. For darker, richer stock, brown the vegetables in olive oil before adding the water. And there's no rule against adding a ham bone, turkey carcass, chicken bones, or bit of leftover beef to the pot.

4 carrots, cut into chunks

2 medium onions or 1 large onion, quartered

2 potatoes, cut into chunks

2 or 3 celery stalks, roughly chopped

3 or 4 cloves of garlic

20 or so stems of parsley, with or without leaves

Salt and freshly ground black pepper

1 Combine everything in a stockpot with 12 cups water, a pinch of salt, and some pepper. Bring to a boil and adjust the heat so the mixture simmers steadily but gently. Cook for about 30 minutes, or until the vegetables are tender. Going longer will only improve the flavor, and a few minutes less won't hurt much, either.

2 Strain, then taste and adjust the seasoning before using, taking care not to oversalt if you're reducing the stock further. Cool before refrigerating or freezing.

Quick Shrimp or Fish Stock: The vegetables in the main recipe become optional; they add complexity but aren't essential. In Step 1, use the shells from about 1 pound of shrimp or the bones and scraps from a pound or so of raw fish (your fishmonger often has these for sale or will give them away free). You won't need to simmer the stock as long to extract good flavor; 10 minutes or so ought to do the trick. Strain, season, and use or store as above.

Roasted Red Peppers

Makes: **4 to 8 servings** Time: **20 to 60 minutes**

It's as easy to make a big batch of roast peppers as a small batch, and since they keep for a few days in the fridge (even longer if you cover them in olive oil), why not? The basic idea is to char the skin—by broiling, roasting, or grilling—so it peels off easily, while developing maximum flavor and perfect tenderness.

You'll get the most smokiness from grilling, of course, but all methods are good. Once the peppers are done, toss them with minced fresh or roasted garlic, a sprinkle of grated Parmesan cheese, a drizzle of balsamic vinegar, a few capers or anchovies, or lots of chopped fresh herbs. They're delicious scrambled with eggs, or as a meaty sandwich filling.

8 red, yellow, or green bell peppers, washed

Salt

Olive oil as needed

1 Heat the oven to 450°F. Or heat the broiler or a grill and put the rack about 4 inches from the heat source. To roast or broil: put the peppers in a foil-lined roasting pan, then roast or broil, turning the peppers as each side browns, until they have darkened and collapsed. The process takes 15 or 20 minutes in the broiler, or up to an hour in the oven. To grill: put the peppers directly over the heat. Grill, turning as each side blackens, until they collapse, about 15 minutes.

2 Wrap the cooked peppers in foil (if you roasted the peppers, use the same foil that lined the pan) and cool them enough to handle. Remove the skin, seeds, and stems (this process is sometimes easier under running water). Don't worry if the peppers fall apart. Serve at room temperature (even if they've been refrigerated), sprinkled with a little salt and a little (or a lot) of olive oil. Store in the refrigerator.

Roasted Red Pepper Sauce: Put the roasted red peppers in a food processor or blender with a few drops of olive oil, stock, white wine, or water—just enough to get the machine working. Sprinkle with salt and pepper or add other ingredients (see the headnote) as you like.

Hybrid Quick Bread

Makes: **4 to 6 servings** Time: **About an hour, largely unattended**

An easy-to-make whole wheat bread with a little white flour for lightness. The covered baking technique ensures the biscuit-like crumb will remain light and fluffy. Like most quick breads, this is best eaten the same day, or toasted the next day.

There are lots of possible variations. Stir in chopped fresh hot pepper, sautéed onion, olives, dried fruit, or nuts; to make this more like cornbread, substitute 1 cup cornmeal for a cup of the whole wheat flour, and add a cup or so of corn kernels if you like. If it's true whole grain bread you're after, see Almost No-Work Whole Grain Bread (page 156) and Easy Whole Grain Flatbread (page 224).

- 1/4 cup olive oil, plus more for the pan
- 2 cups whole wheat flour, plus more if needed
- 1 cup all-purpose white flour, plus more as needed
- 1/2 teaspoon baking powder
- 1 teaspoon baking soda
- 1 1/2 teaspoons salt, preferably coarse or sea salt, plus more for sprinkling
- 3/4 cup yogurt or buttermilk
- 3/4 cup warm water
- 2 tablespoons honey (optional)

1 Heat the oven to 375°F. Grease a cookie sheet or 8-inch square baking pan with about a tablespoon of olive oil. Put the flours, baking powder, soda, and salt in a food processor and turn the machine on. Into the feed tube, pour first the 1/4 cup olive oil, then the yogurt, most of the water, and the honey (if you're using it).

2 Process for a few seconds, until the dough is a well-defined, barely sticky, easy-to-handle ball. If it is too dry, add the remaining water 1 tablespoon at a time and process for 5 or 10 seconds after each addition. If it is too wet (this is unlikely), add 1 or 2 tablespoons of whole wheat flour and process briefly.

3 Form the dough into a round and put it on the cookie sheet or press into the prepared pan, all the way to the edges. Bake for 20 minutes then sprinkle the top with a little coarse salt, and continue baking for another 35 to 40 minutes, until the loaf is firm and a toothpick inserted in the center comes out clean. Let cool completely, then cut the bread into slices or squares and serve or store for up to a day.

Almost No-Work Whole Grain Bread

Makes: **1 standard loaf** Time: **14 to 28 hours, almost completely unattended**

Since I first wrote about Jim Lahey's no-knead bread a couple of years ago, it's become something of a craze with many professionals, and home bakers have been experimenting with their own versions and sharing them on blogs and in books and magazines. In case you missed it: the original recipe creates a free-form white loaf, baked in a pot to create an artisan-style crumb. Lovely, but obviously not whole grain.

Enter this dense, full-flavored loaf, which is even easier. Bakers call this kind of loaf "travel bread." It is not as high as typical sandwich bread, but is used the exact same way. Slice it thinly (and try some toasted); it keeps for days simply wrapped in a tea towel on the counter.

To add whole grains, like cracked wheat, quinoa, or millet, soak ½ cup grain in a small bowl, covered with water, for an hour or so. Drain and add to the dough as described in Step 2.

3 cups whole wheat flour, or use 2 cups plus a combination of other whole grain flours like buckwheat, rye, or cornmeal

½ teaspoon instant yeast

2 teaspoons salt

2 tablespoons olive or vegetable oil

Cornmeal or wheat bran for dusting (optional)

Up to 1 cup chopped nuts, seeds, dried fruit, or proofed whole grains (optional: see headnote)

1 Combine the flour, yeast, and salt in a large bowl. Add 1½ cups water and stir until blended; the dough should be quite wet and sticky but not liquid; add some more water if it seems dry. Cover the bowl with plastic wrap and let it rest in a warm place for at least 12 and up to 24 hours. The dough is ready when its surface is dotted with bubbles. Rising time will be shorter at warmer temperatures, or a bit longer if your kitchen is chilly.

2 Use some of the oil to grease the loaf pan. If you are adding nuts or anything else, fold them into the dough now with your hands or a rubber spatula. Transfer the dough to the loaf pan, and use a rubber spatula gently to settle it in evenly. Brush the top with the remaining oil and sprinkle with cornmeal if you like. Cover with a towel and let rise until doubled, an hour or two depending on the warmth of your kitchen. When it's almost ready, heat the oven to 350°F.

3 Bake the bread until deep golden and hollow-sounding when tapped, about 45 minutes. (An instant-read thermometer should register 200°F when inserted into the center of the loaf.) Immediately turn out of the pan onto a rack and let it cool before slicing.

Fast Whole Grain Bread: Increase the yeast to 1½ teaspoons. Reduce the initial rise to 2 hours and the final rise in the pan to 60 minutes or so. Proceed immediately to Step 3.

Olive Oil Drizzle

Makes: **4 servings (1/2 cup)** Time: **5 minutes**

A superfast sauce you can use on nearly everything: simply cooked vegetables, grains, beans, meat, fish, poultry, even toasted bread. You can substitute wine, beer, sake, stock, or juice for the lemon juice and water, and you might try cooking spices or dried chiles, single spices, or seasoning blends along with the onion or other aromatics in Step 1. Or omit the parsley and finish the sauce with a pinch of stronger herbs like rosemary, thyme, or oregano.

1/4 cup olive oil

1 tablespoon minced onion, garlic, ginger, shallot, scallion, or lemongrass

Salt and freshly ground black pepper

2 tablespoons freshly squeezed lemon juice or mild vinegar, like balsamic or rice vinegar

2 tablespoons chopped fresh parsley, cilantro, or mint leaves

1 Put the oil in a small saucepan over medium heat. When the oil is hot, add the onion, sprinkle with salt and pepper, and cook, stirring occasionally, until it softens, just a minute or two. Turn the heat down if it starts to color.

2 Stir in 2 tablespoons water and the lemon juice; maintain the heat so it bubbles gently for a few seconds but doesn't boil away. Taste, adjust the seasoning, and serve.

Sesame Oil Drizzle: For Asian flavors, replace the olive oil with 2 tablespoons each dark sesame oil and peanut oil; substitute scallions instead of the onions, and soy sauce for the lemon juice. If you like, add a small dried hot chile to the scallions as they cook. Use cilantro for the finishing herb.

Fast Roasted Garlic

Makes: **3 heads** Time: **20 to 40 minutes**

Sweet and creamy, roasted garlic is useful on its own or in soups, sauces, stir-fries, sandwiches, and vegetable purees. I like to double or triple the quantity of oil so I can have it for sautéing, sauces, and stir-fries. Just make sure to refrigerate both garlic and oil; they keep for a few days, but not longer.

To make roasted garlic the traditional, whole-head way, put the ingredients in a small baking dish. Then cover with foil and bake, undisturbed, until soft, at least 40 minutes.

3 whole heads garlic

3 tablespoons or more olive oil

Salt

1 Heat the oven to 375°F. Break the garlic heads into individual cloves, but don't peel them. Spread them out in a pan, drizzle with oil, and sprinkle with salt. Bake, shaking the pan occasionally, until tender, about 20 minutes.

2 Let cool enough to handle. To use, squeeze the garlic from the skins or carefully remove the skins with a knife.

Fast Braised Garlic: Increase the olive oil to ⅓ cup and put it in a skillet large enough to hold the garlic in one layer, over medium-low heat. When the oil is hot, add the garlic. Sprinkle with salt. Adjust the heat so the garlic barely sizzles. Cook, turning occasionally so the garlic browns evenly, until it gradually turns golden, then begins to brown. The garlic is done when perfectly tender; it should take about 15 to 20 minutes. Store as above.

Breakfast

You've heard all your life that breakfast is the most important meal of the day. I'm not prepared to either verify or deny that platitude, but I will say this: A sturdy breakfast will certainly help you set the tone for the rest of your day's eating, and steer the midmorning snack away from doughnuts.

I'm a fan of savory breakfasts, as are the citizens of much of the world, so what turns me on is a bowl of beans and rice, or leftover soup from the night before, or even a dish of lightly sautéed vegetables. And if you think about it, much of what we consider the traditional American breakfast—eggs, bacon, sausage, toast, potatoes—is quite savory (and would serve as a fairly large supper). If you're looking for strong, mostly salty (and undeniably satisfying) flavors at breakfast, try the frittata on page 170, and don't miss the list of Other Dishes in the Book You Can Eat for Breakfast or Brunch, page 175.

If you prefer sweet breakfasts, there's no reason you can't be happy in the Food Matters framework. Have a smoothie or a big bowl of fruit salad, seasoned with lemon juice or mint, with maybe a spoonful or two of yogurt if you like. Or have a bowl of granola or cooked whole grains (you can cook a batch in advance and reheat a bit at a time each morning, in the microwave) with, if you like, a splash of milk or maple syrup and even a bit of butter.

A Very Flexible Fruit Salad

Makes: **4 servings (about 6 cups)** Time: **30 minutes**

It isn't always easy—especially in the dead of winter—to find perfect fruit, which is what it takes to make an ideal fruit salad. But choosing what's ripe and fresh may mean opting for less traditional combinations, and that can be fun. Choose berries and apricots in early summer; tomatoes (yes!) and peaches with basil later on. Apples and pears become staples in the fall, and you can augment them with dried fruit.

To tide you through the winter, turn to avocado and pineapple (maybe with chopped cilantro and a good squeeze of lime; a pinch of chili isn't bad, either), or a mixture of citrus and bananas.

Try combining chunks of different melons, sprinkled with sea salt and chopped mint, or top sliced pears with hazelnuts, nuts, or crumbled blue cheese. In every case, figure about 1½ cups per person.

½ medium cantaloupe or honeydew, seeded, peeled, and cut into chunks

1 mango or papaya (or the other half of the melon) seeded, peeled, and cut into chunks

¼ pineapple, peeled, cored, and cut into chunks

Zest and juice of 1 lemon

1 pint strawberries, hulled and halved

1 pint other berries (blackberries, blueberries, raspberries, etc.)

1 orange or other citrus fruit, peeled and segmented

1 Put the melon, mango, pineapple, and lemon zest and juice in a large bowl and toss; add the remaining ingredients and toss gently, taking care not to crush the berries.

2 Serve immediately, or refrigerate and serve within a few hours of mixing.

Fruit Smoothies

Makes: **2 to 4 servings (about 4 cups)** Time: **5 minutes**

Smoothies can range from simple to elaborate mixtures, but they're an almost instant breakfast, lunch, snack, or dessert. You can start with fresh or frozen fruit (unsweetened frozen fruit is convenient, inexpensive, and sometimes more flavorful than fresh). Smoothies work with or without milk or yogurt or additional sweetening, depending on just how fruity you want them. I use frozen banana in many of my smoothies, for flavor and texture.

Unless the fruit you're using is very watery, you'll need some extra liquid to get a nice puree. The recipe here gives you some options and starting proportions. If you go the fresh fruit route, add some ice cubes for frostiness. If you want to skip the dairy ingredients altogether, omit the yogurt, use the whole frozen banana, and increase the fruit juice as needed for the right consistency. To vary the flavors, try coconut milk, pineapple juice, or exotic nectars.

3 cups unsweetened frozen or fresh fruit, any combination (strawberries, blueberries, mango, peaches, melon, etc.)

½ frozen banana

1 cup plain yogurt, nondairy milk, or juice

½ cup apple juice or other liquid; more if needed

Maple syrup or honey to taste (optional)

Put all ingredients in a blender and whiz until smooth. If the machine is not puréeing, add a little more apple juice or other liquid as needed. Serve immediately.

Porridge, Updated

Makes: **4 to 6 servings** Time: **15 minutes**

Porridge is an ancient and international breakfast staple, for good reason: It's cheap, easy, and nutritious. Oatmeal and cornmeal mush are the American classics, but you need not stop there: Use any ground, cut, or rolled grain. Steer clear, though, of fast-cooking or instant oats, which are tasteless. For extra flavor or texture, stir in fresh or dried fruit, nuts and seeds, vanilla, or ground spices, like cinnamon, nutmeg, cloves, or cardamom; add sweetness with a little honey or syrup or richness with a small pat of butter, or a spoonful of milk or cream.

To go savory, try topping the porridge with a spoonful of chunky salsa, grated cheese, chopped hard-boiled egg (or, for a special treat, poached or fried egg), a drizzle of soy sauce and a few sliced scallions, or simply coarse salt and freshly ground pepper. Or fold in leftover chopped vegetables (mushrooms are nice) or raw tender greens (like spinach) and let them wilt a bit.

Cooking times will vary depending on the grain you use, but even if you're cooking for yourself you may as well make a full batch, since this keeps for days in the fridge and reheats perfectly in the microwave. (This means you can make porridge at night or any other time that's convenient for you.)

Pinch of salt

2 cups grain, like rolled oats (or other rolled grain), cornmeal (or grits), cracked wheat, quinoa, millet, or short-grain brown rice

Butter, to taste (optional)

Salt, sweetener (like maple syrup, sugar, or honey), and/or milk or cream, as desired

1 Combine 4 to 4½ cups water (more water will produce creamier porridge), the salt, and the grain or grains in a medium saucepan and turn the heat to high. When the water boils, turn the heat to low and cook, stirring frequently, until the water is just absorbed: about 5 minutes for rolled oats, 15 minutes for cornmeal or cracked wheat, 30 minutes for quinoa or millet, or up to 45 minutes or more for brown rice. Add water as needed to keep the porridge from sticking.

2 When the grains are very soft and the mixture is thickened, serve or cover the pan and turn off the heat; you can let sit for up to 15 minutes. Uncover, stir, add other ingredients as desired, and serve.

Breakfast Couscous

Makes: **4 servings** Time: **20 minutes**

Two of my favorite breakfast grains are couscous and bulgur (which is still sold as the commercial breakfast cereal Wheatena; see the variation). Neither requires cooking (they both steep, like tea), so they're perfect for even the busiest mornings—and really tough to screw up.

Both take perfectly to fresh or dried fruit, nuts, and a drizzle of maple syrup or honey; milk is optional. You can also take this in a savory direction, adding leftover sautéed mushrooms along with bits of sausage, chopped ham, or bacon. See Leftover Grains for Breakfast, on page 169, for specific ideas.

- 1 cup whole wheat or white couscous
- Salt
- 1 cup fresh fruit (sliced bananas, berries, diced apples, peaches)
- 1/4 cup chopped nuts (optional)
- 1/4 cup dried fruit like raisins, dates, or coconut (optional)
- Drizzle of honey or maple syrup

Put the couscous in a medium pot with a tight-fitting lid and add 1½ cups of water and a pinch of salt. Bring the water to a boil, then cover and remove from the heat. Let steep for at least 10 minutes (5 minutes if using white couscous), or up to 20. Add the fruit, nuts, and honey if using. Fluff with a fork and serve.

Breakfast Bulgur: Use 1 cup of any grind of bulgur, with 2½ cups boiling water. Stir once and let sit. Fine bulgur will be tender in 10 to 15 minutes, medium in 15 to 20 minutes, and coarse in 20 to 25. If any water remains when the bulgur is done, put it in a fine strainer and press down on it, or squeeze the bulgur in a cloth. Fluff with a fork and add fruit, nuts, and honey, if using.

Alternatives to Milk

Reducing consumption of animal protein sometimes means looking beyond ordinary milk, and there are good alternatives, made from nuts, grains, and legumes. Just be sure to read the labels. Most are sweetened; look for packages that specifically say "unsweetened." Many are flavored with vanilla or even chocolate (which you might like as long as you're not expecting plain). And some, especially nut milks, include gums or other ingredients.

Soy milk: Almost as high in protein as cow's milk, soy milk makes a fine daily alternative for coffee, tea, and cereal. It's also handy because it separates less during heating than other milk alternatives; this sometimes makes it a good substitute for baking and cooking.

Nut milk: A great choice for desserts, grain dishes, and thick soups; since it actually adds a welcome flavor to any dish that takes to the taste of nuts. Also really nice as a replacement for cream or half-and-half in coffee.

Oat milk: With a consistency similar to low-fat or skim milk, oat milk is good for drinking but a little thin for cooking. It has a neutral taste and a pretty golden color.

Rice milk: Slightly sweeter than oat milk or soy milk, this has a neutral flavor and a thin, almost watery consistency.

Coconut milk: With a lovely flavor and a thick consistency, coconut milk is ideal for desserts and Asian soups, stews, and sauces; it heats up beautifully. Though it's quite high in fat, light or reduced-fat coconut milk is usually a fine substitute.

Anything Goes Granola

Makes: **About 9 cups** Time: **30 minutes**

Granola is a versatile snack and breakfast food. Unfortunately, most packaged stuff is usually too sweet, more like candy than cereal. Though some of the smaller production brands are nice, it remains worth making and customizing your own.

Usually, granola has a high proportion of rolled oats, but you can add different rolled grains like wheat, rye, or kamut. Play with the flavor by tossing different nuts and seeds into the mix, adding a teaspoon of vanilla or ground spices like cinnamon, ginger, cardamom, or nutmeg (alone or in combination), or using any chopped dried fruit—dates, cranberries, cherries, blueberries, apricots, pineapple, crystallized ginger, or banana chips.

5 cups rolled oats (*not* quick-cooking or instant) or other rolled grains

3 cups mixed nuts and seeds, like sunflower seeds, chopped walnuts, pecans, almonds, cashews, and sesame seeds

1 cup shredded, unsweetened coconut

1 teaspoon ground cinnamon, or other spices to taste

½ to 1 cup honey or maple syrup, or to taste

Salt

1 teaspoon vanilla (optional)

1 to 1½ cups raisins or chopped dried fruit

1 Heat the oven to 350°F. In a large bowl, combine the oats, nuts and seeds, coconut, cinnamon, sweetener, and vanilla if using; sprinkle with a little salt. Toss well to thoroughly distribute ingredients. Spread the mixture on a rimmed baking sheet and bake for 30 minutes or a little longer, stirring occasionally. The granola should brown evenly; the darker it gets without burning, the crunchier it will be.

2 Remove pan from oven and add raisins. Cool on a rack, stirring now and then until the granola reaches room temperature. Put in a sealed container and store in refrigerator; it will keep indefinitely.

Swiss-Style Muesli: Basically uncooked granola, so it's even easier: Omit the vanilla and honey or syrup. Combine the oats, nuts, seeds, coconut, cinnamon, and raisins in a large bowl, and sprinkle with salt. Toss the mixture with ¼ cup brown sugar. Serve with yogurt, fresh fruit, honey, or milk. Store as you would granola.

Leftover Grains for Breakfast

Freshly made oatmeal, couscous, and bulgur are all lovely ways to start the day. But since you're integrating whole grains into your daily diet (easy enough, now that you're cooking Whole Grains without Measuring on page 136), there are probably leftovers in the fridge.

A reheated bowl of any grain—brown rice, quinoa, barley, millet, cracked wheat, farro, or wheat berries—quickly becomes breakfast, with the addition of some fresh or dried fruit, nuts, or ground spices. (The microwave is ideal for heating leftover grains, or put them in a small pot with a few drops of water and set on the stove over low heat.) Add a splash of milk, maple syrup, or honey if you like. Or go the savory route and toss the grains with a spoonful of leftover sautéed veggies, chunky salsa, or a drizzle of soy sauce and handful of sprouts.

More-Vegetable-Than-Egg Frittata

Makes: **2 or 4 servings** Time: **30 minutes**

Frittata is great anytime, hot or at room temperature, and can be made with almost anything. In the Food Matters kitchen, the ratio of vegetable to egg changes dramatically, with terrific results: I use four to six cups of vegetables and just two or three eggs for two hungry people, or for four or more servings as part of a larger meal or a larger appetizer. The vegetables remain dominant and delicious.

You can start with either cooked or raw vegetables: Try ribbons of spinach or chard, chopped fresh or dried tomatoes, potato slices, asparagus, broccoli rabe, sautéed mushrooms, zucchini, or eggplant cubes. Fresh basil is lovely with nearly everything, but other herbs like tarragon or mint are also super. And of course you can toss in some cooked crumbled sausage, bacon, or chopped ham, or even shrimp just before adding the eggs.

2 tablespoons olive oil

½ onion, peeled and sliced

Salt and freshly ground black pepper

4 to 6 cups of any chopped or sliced raw or cooked vegetables, drained of excess moisture if necessary

¼ cup fresh basil leaves, or 1 teaspoon chopped fresh tarragon or mint leaves (optional)

2 or 3 eggs

½ cup freshly grated Parmesan cheese, optional

1 Put a tablespoon of the olive oil in a skillet and turn the heat to medium. When the oil is hot, add the onion and cook, sprinkling with salt and pepper, until it's soft, about 3 minutes. Add the vegetables, raise the heat, and cook, stirring occasionally, until they soften, anywhere from a couple of minutes for greens to 15 minutes for sliced potatoes. Adjust the heat so the vegetables brown a little without scorching. (If you're starting with precooked vegetables, add them to the onions and give a couple of good stirs before proceeding.)

2 When the vegetables are nearly done, turn the heat to low and add the basil. Cook, stirring occasionally, until the pan is almost dry, up to another 5 minutes for wetter ingredients like tomatoes or mushrooms.

3 Meanwhile, beat the eggs with some salt and pepper, along with the cheese if you're using it. Pour over the vegetables, using a spoon if necessary to distribute them evenly. Cook, undisturbed, until the eggs are barely set, 10 minutes or so. (You can set them further by putting the pan in a 350°F oven for a few minutes, or running it under the broiler for a minute or two.) Cut into wedges and serve hot, warm, or at room temperature.

Better Poached Eggs: Soupy in a good way. In Step 2, instead of cooking the vegetables until dry, when they're still soupy and on the raw side, add 2 cups (or more) vegetable stock or water. Bring to a gentle boil and carefully crack the eggs into the bubbling mixture. Cook, uncovered, until the eggs are set and done as you like them, anywhere from 3 to 7 minutes. Scoop the eggs, the vegetables, and some of the cooking liquid into bowls and serve.

Whole Grain Pancakes

Makes: **About 6 servings** Time: **30 to 40 minutes**

The secret to light whole grain pancakes is to beat the egg whites really well, so the batter can support not only *all* whole grain flour—no mean feat—but small amounts of add-ins as well. Some ideas to get you started: ½ cup cornmeal, rolled oats, or oat or wheat bran in place of ½ cup of the flour; up to 2 tablespoons of ground flaxseed; add up to ½ cup any light, cooked grains like couscous, millet, or quinoa; freshly grated orange or lemon zest; chopped nuts; berries or sliced bananas; unsweetened shredded coconut or chopped dried fruit.

For an exotic bread replacement, omit the sugar, increase the salt a bit, replace the cinnamon with cumin, and serve the pancakes as flatbreads with soups, stews, or salads.

Butter as needed

1⅔ cup whole wheat flour

2 tablespoons sugar

1 tablespoon baking powder

½ teaspoon ground coriander or cardamom (optional)

½ teaspoon ground cinnamon (optional)

½ teaspoon salt

2 large eggs, separated

2 cups milk

1 Melt 3 tablespoons butter. In a large bowl combine flour, sugar, baking powder, spices, and salt.

2 Beat the egg whites with an electric mixer or a whisk until stiff peaks form, but do not overbeat. In a separate bowl beat the yolks, milk, and melted butter until foamy, a couple of minutes. Add the milk mixture to the flour mixture and give a couple of good stirs, but do not overmix. Fold in the egg whites and stir until the batter is just evenly colored and relatively smooth; it's OK if there are some lumps.

3 Heat a large skillet (preferably cast-iron) or griddle over medium heat until a few drops of water dance on its surface. Add the butter as needed (or use a thin film of vegetable oil). When the skillet is hot, spoon the batter into pan. Cook until bubbles form and pop, about 2 minutes; you may have to rotate the cakes to cook them evenly, depending on your heat source and pan. Then carefully flip pancakes. Cook until well colored on other side, another minute or two more. Serve or keep in warm oven for a few minutes. Serve with maple syrup, fruit compote, jam, or caramelized apples.

Breakfast Bread Pudding

Makes: **4 to 6 servings** Time: **About 1½ hours, largely unattended**

Not your usual bread pudding; this has less custard and more bread, fruit, and nuts. For variety, use pears, peaches, cherries, or blueberries instead of the apples. Or go savory (see the variation).

- Butter or grape seed oil or other oil for greasing the pan
- 2 eggs
- 1 cup milk
- ¼ cup honey, or to taste
- 1 teaspoon ground cinnamon
- Pinch salt
- 4 medium to large apples, cored, peeled (or not), and cut into chunks or slices
- ½ cup raisins (optional)
- ½ cup chopped walnuts or hazelnuts (optional)
- 8 slices whole or multigrain bread (preferably stale), cut in 1-inch cubes (about 3 cups)

1 Heat the oven to 350°F. Butter a 1½-quart or 8-inch-square baking dish. Beat the eggs in a large bowl. Whisk in the milk, honey, cinnamon, and salt. Stir in the apples, raisins, and nuts. Then fold in the bread cubes, using your hands or a rubber spatula to make sure everything is evenly coated. Let the mixture sit for about 15 minutes or until all of the liquid has been absorbed; give another good stir. (You can prepare the pudding ahead to this point; cover and refrigerate for up to 12 hours.)

2 Transfer the mixture to the prepared dish and smooth out the top. Bake for 40 to 45 minutes or until golden and only a little wobbly in the center. Let sit for a few minutes before cutting.

Serve warm or cold. This keeps well for 2 days or more, covered and refrigerated.

Savory Bread Pudding: Use olive oil to grease the pan if you like. Omit the honey, cinnamon, apples, raisins, and nuts. Instead use 4 or 5 cups lightly cooked vegetables, like artichoke hearts, asparagus, potatoes, eggplant, mushrooms, sautéed spinach, or roasted tomatoes. Proceed with the recipe, reducing the baking time to 30 to 40 minutes.

Other Dishes in the Book You Can Eat for Breakfast or Brunch

Boiled or Steamed Vegetables, As You Like 'Em (page 132)

Hybrid Quick Bread (page 154)

Almost No-Work Whole Grain Bread (page 156)

Whole Grain Bread Salad with dried fruit (page 192)

Impromptu Fried Rice (page 208)

Vegetable Spread, from just about any vegetable (page 222)

Vegetable Pancakes (page 230)

Breakfast Burritos

Makes: **4 servings** Time: **15 minutes with precooked (or canned) beans and rice**

With cooked beans in the fridge, breakfast burritos become a quick (and portable!) meal. You can always add some scrambled egg or cheese, but I use these wraps as my vehicle for whatever leftover vegetables or greens are handy. Sometimes I add rice or potatoes, but again this depends on what remains from last night's dinner. Finish with the usual suspects: salsa or chopped fresh tomatoes are a must, but cilantro, black olives, avocado, and fresh or pickled chiles all work, too.

To make life even easier, double or triple the batch (omitting the lettuce), wrap the burritos well in foil, and freeze. On a busy morning, remove the foil, wrap the burrito in a paper towel and zap it for a couple of minutes in the microwave. Or reheat foil-wrapped burritos in a 350°F oven for about 20 minutes.

2 cups plain cooked beans (Pot of Beans, page 139), or Bean and Vegetable Chili (page 260), or canned pinto or black beans

1 cup cooked brown rice or potatoes (optional)

4 large whole wheat flour tortillas

2 to 4 cups salad greens, torn-up lettuce, or cooked leftover vegetables (or a combination)

2 cups fresh or cooked salsa, or chopped fresh tomatoes

1 Warm the beans or chile with the rice, if using, in a small pot or in the microwave. To warm the tortillas, wrap them in foil and put in a 300°F oven for about 10 minutes, or stack them between two damp paper towels and microwave for 30 to 60 seconds.

2 Put a tortilla on a plate or flat surface and put the bean mixture toward the bottom. Top with the greens, then the salsa and any other ingredients. Fold both sides over to enclose the filling, then roll up, and serve.

Lunch

Lunch is the meal most likely to be eaten out of the home, so it presents its own specific challenges. A salad bar, a row of cooked cold vegetables and grains, room temperature grilled vegetables, even rice and beans, all are preferable to the company cafeteria steam table or a sandwich or grill station.

The best bet is often to bring your own lunch. And this is a solution that may be easier than you think, as long as you're equipped with the right containers. The most convenient for all-around use are tight-sealing plastic (or better still, glass) so you can either dress your food at home, or carry dressings and sauces separately. If you have access to a microwave at work, you have even more options.

Here are recipes for dozens of different brown-bag lunches: soups, salads, sandwiches, and even noodle and stir-fry dishes. Personally, I bring lunch to work about half the time, and it's usually a soup (or something soup-like if not an actual "soup"), one that contains vegetables or legumes and grains, so it's pretty substantial.

They're based on the same quickly prepared foods you would eat if you were home. Any of them can be made in the morning and reheated if necessary; most can even be made the night before. If you prepare just one or two of these every week (doubling them if you're feeding a family this way), you have lunches all set, especially if you add to the bounty by bringing leftovers from dinner once in a while.

Chopped Cabbage Salad

Makes: **6 servings** Time: **10 minutes, plus 1 to 2 hours for salting**

You might call this a slaw, though the result is much more like a salad, especially if you take the time to salt the cabbage beforehand, which makes it tender yet still crunchy. The other advantage to not using lettuce is that, like coleslaw, this salad keeps for a couple of days and gets even better over time.

1 small head cabbage (about 1 pound), any kind

Salt as needed

2 celery stalks, chopped

2 carrots, chopped

1 small to medium red onion, minced

1 red or yellow bell pepper, cored, seeded, and chopped

⅓ cup olive oil

2 tablespoons sherry or white wine vinegar, or lemon juice

Freshly ground black pepper

Chopped fresh parsley leaves for garnish, optional

1 Core the cabbage and chop it roughly. If you're not salting, skip to Step 2. If you are, put the cabbage in a colander, sprinkle with about 2 teaspoons of salt, and toss. Check in 10 minutes or so to see if the leaves are exuding moisture. If not, add a little more salt and toss again. Let sit an hour or two, pressing the moisture out with your hands once or twice. Taste it; if it's too salty, rinse and pat dry.

2 Combine the vegetables in a bowl with the cabbage; sprinkle lightly with pepper (and salt if you didn't salt the cabbage), add the olive oil and vinegar, and toss. Taste and adjust the seasoning, garnish if you like, and serve.

Chopped Cabbage Salad, Asian Style: Instead of the olive oil, use 1 tablespoon sesame oil and 5 tablespoons peanut oil. Substitute lime juice or rice vinegar for the sherry vinegar or lemon juice. Add some minced fresh hot chile and chopped scallions if you like, and use chopped cilantro leaves to garnish instead of the parsley.

Nice Additions to Chopped Cabbage Salad

1 peeled, cubed avocado

Any green beans, about 1 cup, cooked briefly and shocked (page 132)

Fennel, 1/2 bulb or so, trimmed and chopped (in place of the celery)

Radishes, 1/2 cup chopped

Fresh peas, snow peas, or snap peas, about 1 cup, very lightly cooked and shocked (page 132)

New potatoes, steamed and cut into small chunks—about 1 cup

Chickpeas, 1/2 cup (or more) cooked or canned, drained

Crumbled blue or feta cheese (add to the main recipe only, *not* to the variation)

Shredded crab, or chopped shrimp or chicken

Layered Salad

Makes: **4 servings** Time: **20 minutes**

For 20 years we've served alternating slices of tomatoes and mozzarella as if they were the only ingredients available for layering in a salad. It's a fine combination, to be sure, but come on—we can branch out a bit.

This version provides a formula for all sorts of combinations to try all year long, even when tomatoes aren't in season; just pick one of the options from each line in the ingredient list and build away.

- 3 large ripe tomatoes, cored; or 3 oranges (blood oranges are nice), peeled
- Salt and freshly ground black pepper
- 1 cucumber, peeled, seeded, and sliced; or 8 thick slices of watermelon (no bigger than the tomatoes), rinds and seeds removed
- 1 small red onion, halved and sliced paper-thin; or 2 scallions, sliced thinly
- 1 medium avocado, pitted, peeled, and sliced; or 8 thin slices fresh mozzarella or feta cheese
- 1/4 pound jicama, daikon, or other radishes, Asian pear, or Granny Smith apple, peeled if necessary, grated or finely chopped (about 3/4 cup)
- 2 tablespoons olive oil
- 1 tablespoon freshly squeezed lemon or rice or sherry vinegar
- 1/2 cup chopped fresh basil or cilantro for garnish

1 Cut each tomato or orange crosswise into 4 thick slices. Put a layer of them on plates or a platter and sprinkle with salt and pepper. Top with the cucumber or watermelon, then the onion or scallions; season again if you like. Put the avocado or cheese slices over all, fanning them out to get good coverage. Top with the radishes or apple and sprinkle with salt and lots of pepper one last time.

2 Use a fork to mix the oil and lemon juice or vinegar in a small bowl and drizzle the dressing over the top. Garnish with the chopped basil or cilantro and serve.

Layered Peach Salad: A must when they're in season. Cut 2 peaches into thin slices and layer them into the salad after the avocado or cheese and before the radishes or apples. Dress and garnish as above.

Lunchtime Caesar Salad, Revisited

Makes: **4 servings** Time: **20 minutes**

It takes just a couple of tweaks to convert the classic—and not particularly nutritious—Caesar salad into Food Matters fare. The meat is optional here; you could also toss the salad with a cup or so of cannellini beans. Lightly cooked vegetables, like asparagus, string beans, or mushrooms, are all welcome additions to, or substitutions for, the zucchini.

- 1 or 2 chicken breasts, or ½ pound peeled shrimp (optional)
- Salt and freshly ground black pepper
- 5 to 6 tablespoons olive oil
- 1 clove garlic, halved
- 2 eggs or ⅓ cup soft silken tofu
- 2 tablespoons freshly squeezed lemon juice
- 1 or 2 minced anchovies, or to taste (optional)
- Dash of Worcestershire sauce
- 1 medium zucchini or 2 small zucchinis, grated and squeezed dry in a clean towel
- 1 large head romaine lettuce, torn into pieces
- Croutons (page 234)
- ½ cup freshly grated Parmesan cheese

1 If you're serving the salad with chicken or shrimp, heat a charcoal or gas grill or the broiler and adjust the rack about 4 inches from the heat source. Sprinkle the chicken or shrimp with salt and pepper and coat the pieces with a tablespoon of the olive oil. When the grill or broiler is hot, grill the chicken or shrimp, turning as needed, until browned outside and just opaque inside, 6 to 8 minutes for the chicken or about 3 minutes for the shrimp. Remove from the heat and set aside. (If you're not using chicken or shrimp, skip to Step 2.)

2 Rub the inside of a large salad bowl with the garlic clove, then discard the clove. Bring a small pot of water to a boil and cook the eggs in the boiling water for 60 to 90 seconds; they will just begin to firm up. Crack them into the salad bowl, being sure to scoop out the white that clings to the shell. (If you're using the tofu, simply put it into the bowl.)

3 Whisk the eggs or tofu, gradually adding the lemon juice and then the olive oil, beating all the while. Whisk in the anchovies if you're using them (mashing them a bit), and the Worcestershire sauce. Taste and add salt if needed and plenty of pepper. Toss well with the zucchini and lettuce; top with the croutons, Parmesan, and chicken or shrimp if using it, then toss again at the table and serve.

Salade Niçoise with Mustard Vinaigrette

Makes: **4 servings** Time: **30 minutes**

I've always considered the tuna in this classic French country salad optional, since the combination of green beans, potatoes, and olives is quite satisfying, but add whatever ingredients you like from the list that follows. If you have leftovers to add, especially beans and potatoes, all the better.

The quick mustard vinaigrette is definitely garlicky; simply omit the garlic if that's not your thing. And since you can always use this dressing for other dishes, the recipe makes enough for you to have some left over; it keeps in the fridge for a couple of days.

¼ cup red wine vinegar, plus a little more if needed

⅔ cup olive oil, more or less

2 garlic cloves, minced

2 teaspoons Dijon mustard

¼ cup fresh chopped herbs like chervil, dill, tarragon, or parsley

Salt and freshly ground black pepper

½ pound fresh green beans, trimmed

½ pound small waxy potatoes, like red skin or fingerlings, scrubbed but not peeled, and cut in half

8 cups torn assorted lettuces and other salad greens, including some whole basil leaves if you like

1 cup good black olives, preferably niçoise or any oil-cured

4 ripe tomatoes, cored, seeded, and cut into wedges

1 small red onion, halved and thinly sliced

1 Set a large pot of water to boil and salt it; get a bowl of ice water ready by the stove. Make the vinaigrette by putting the vinegar, oil, garlic, and mustard in a glass jar or small bowl;

Some Possible Additions to Salade Niçoise

2 hard-boiled eggs, sliced or cut into wedges

1 can tuna, preferably packed in olive oil, or 8 ounces grilled or broiled fresh tuna, cut into chunks

4 to 8 anchovies

1 cup large lima or cannellini beans, or chickpeas

Sliced or chopped cucumbers

Sliced radishes

Sliced red or yellow bell pepper

Leftover roasted vegetables like leeks, eggplant, carrots, or parsnips

sprinkle with salt and lots of pepper and add the herbs if you're using them. Shake or stir and taste. Add more vinegar or oil if necessary and adjust the seasoning. Set aside.

2 When the water boils, put the green beans into the pot. When they turn bright green and are crisp-tender (in about 3 minutes), use tongs or a slotted spoon to fish them out and plunge them into the ice water to cool. Drain the beans and refill the bowl with ice water.

3 Add the potatoes to the pot. Adjust the heat so the mixture bubbles enthusiastically. Start checking the potatoes in about 5 minutes. You want them tender at the core, but not falling apart. When they're done, plunge them into the bowl of ice water to cool; then drain.

4 Arrange all the salad ingredients nicely on a platter: greens on the bottom, topped with the beans, potatoes, tomatoes, onions, and olives, plus any other ingredients you like. Drizzle with half of the vinaigrette. Or for a more rustic dish, simply toss all the ingredients together. Pass the remaining vinaigrette at the table, or reserve it for later.

Thai Beef Salad

Makes: **4 servings** Time: **25 minutes**

Classic and amazing, this is a dish I crave all the time, with just the right balance of meat to greens. Shrimp and squid are other good options (reduce the cooking time by half). It's also easy enough to take the salad in a vegetarian direction, either by substituting tofu slices (the recipe otherwise remains the same), or by adding 2 cups of cooked edamame or other beans when you toss the salad. Note that this is a perfect use for leftover grilled beef.

8 ounces skirt or flank steak, leftover or raw

6 cups torn salad greens (mixed greens are nice)

1 cup torn fresh herb leaves (mint, cilantro, Thai basil, or a combination)

1/4 cup minced red onion

1 medium cucumber, peeled if necessary, cut in half lengthwise, seeded, and diced

1 small fresh hot red chile, like Thai, or to taste, minced

Juice of 2 limes

1 tablespoon sesame oil

1 tablespoon fish sauce (nam pla, available at Asian markets) or soy sauce

1/2 teaspoon sugar

1 If you are beginning with raw meat, start a gas or charcoal grill or heat a broiler; the rack should be about 4 inches from the heat source. Grill or broil the beef until medium-rare, turning once or twice, 5 to 10 minutes, depending on the thickness; set it aside to cool.

2 Toss the lettuce with the herbs, onion, and cucumber. Combine all of the remaining ingredients with 1 tablespoon of water (the dressing will be thin) and use half of this mixture to toss the greens. Remove the greens to a platter, reserving the dressing.

3 Slice the beef thinly, reserving its juice; combine the juice with the remaining dressing. Lay the slices of beef over the salad, drizzle the dressing over all, and serve.

Tabbouleh, My Way

Makes: **4 servings** Time: **40 minutes**

I used to say that tabbouleh was all about the herbs, and they're still crucial; now, however, I make this salad with more than the traditional amount of vegetables. The result is a flavorful, hearty chopped salad.

You can substitute any small cooked grain like quinoa or steel-cut oats for the bulgur. If you can finesse the timing, dress while still warm, so the grain absorbs all the flavors. If you want to substitute leftover cooked vegetables for the raw ones here (roasted root veggies are quite good, and artichoke hearts are out of this world), figure on about 2 cups. You can also add sesame or sunflower seeds; chopped pistachios or almonds; a handful of white beans or chickpeas; bits of cooked shrimp or squid; or small cubes of tofu or feta cheese are excellent, too.

½ cup fine-grind (#1) or medium-grind (#2) bulgur

⅓ cup olive oil, or more as needed

¼ cup freshly squeezed lemon juice, or to taste

Salt and freshly ground black pepper

1 cup roughly chopped parsley leaves

1 cup roughly chopped fresh mint leaves

1 cup peas or fava beans (frozen are fine; just run them under cold water to thaw)

6 or 7 radishes, chopped

½ cup chopped scallions

2 medium tomatoes, chopped

About 6 black olives, pitted and chopped, or more to taste (optional)

1 Soak the bulgur in 1¼ cups boiling water to cover until tender, 10 to 20 minutes, depending on the grind. If any water remains when the bulgur is done, put the bulgur in a fine strainer and press down on it, or squeeze it in a cloth. Toss the bulgur with the oil and lemon juice and sprinkle with salt and pepper. (You can make the bulgur up to a day in advance. Cover and refrigerate; bring to room temperature before proceeding.)

2 Just before you're ready to eat, add the remaining ingredients and toss gently; taste, adjust the seasoning, adding more oil or lemon juice as needed, and serve.

Whole Grain Bread Salad

Makes: **4 servings** Time: **45 minutes**

I use cherry tomatoes for this hearty *panzanella* because they add chew without a lot of unwanted moisture. But chopped ripe tomatoes—especially if they're perfectly ripe—are equally wonderful. (When good tomatoes aren't in season, try the variation.) Grilled or roasted vegetables, especially eggplants, make a fine addition. Finally, these recipes can double as stuffing for meat, poultry, or vegetables, or can simply be baked in a pan until crusty.

- 8 ounces whole grain bread (4 thick slices; stale is fine)
- Salt
- 8 ounces escarole or lacinato kale (also called black kale or *cavalo nero*), or other kale, or collards
- 1/4 cup olive oil
- 2 tablespoons balsamic vinegar
- 1 1/2 pounds cherry tomatoes, halved
- 1/2 red onion, thinly sliced
- Freshly ground black pepper
- 1/2 cup fresh chopped basil leaves

1 Heat the oven to 400°F. Put the bread on a baking sheet and toast, turning once or twice, until golden and dry, 10 to 20 minutes, depending on the thickness of the slices. Remove from the oven and cool.

2 While bread toasts, put a large pot of water to boil and salt it. Fill a large bowl with ice water. When the pot comes to a boil, add the greens; let wilt for about a minute, then drain and plunge into ice water. When cool, squeeze dry and roughly chop. Put the greens, oil, vinegar, tomatoes, and onion in a large salad bowl. Sprinkle with salt and lots of pepper and toss to coat.

3 Fill a bowl with tap water and soak the bread for about 3 minutes. Gently squeeze a slice dry and crumble it into salad bowl. Repeat with remaining slices. Toss well to combine the salad and let sit for 15 to 20 minutes. Right before serving, taste and adjust seasoning and toss with the basil.

Whole Grain Bread Salad with Dried Fruit: Use any kind of kale or collards instead of the escarole. Omit the tomatoes and basil and substitute shallot for the onion if you like. In Step 2, toss the kale and dressing with 1 cup chopped dried fruit (figs, dates, apricots, cherries, cranberries, or raisins are all good) and 1 tablespoon chopped fresh sage leaves. Garnish with toasted hazelnuts if you like.

Spinach and Sweet Potato Salad with Warm Bacon Dressing

Makes: **4 servings** Time: **About 45 minutes**

A little bacon goes a long way here, and adding roasted sweet potatoes (or any roasted root vegetable, including waxy potatoes) turns an ordinary spinach salad into a perfect one-dish lunch or supper.

2 large sweet potatoes, peeled and cut into bite-size pieces

1/4 cup olive oil

Salt and freshly ground black pepper

2 thick slices of bacon

1 red bell pepper, cored and chopped

1 small red onion, halved and thinly sliced

1 tablespoon peeled, minced fresh ginger

1 teaspoon ground cumin

Juice from 1 orange

1 pound fresh spinach leaves

1 Heat the oven to 400°F. Put the sweet potatoes on a baking sheet, drizzle with 2 tablespoons of the oil, sprinkle with salt and pepper, and toss to coat. Roast, turning occasionally, until crisp and brown outside and just tender inside, about 30 minutes. Remove and keep them on the pan until ready to use.

2 While the potatoes cook, put the bacon in a nonreactive skillet and turn the heat to medium. Cook, turning once or twice, until crisp. Drain on paper towels and pour off the fat, leaving any darkened bits behind in the pan. Put back on medium heat, and add the remaining oil to the pan. When it's hot, add the bell pepper, onion, and ginger to the pan. Cook, stirring once or twice, until no longer raw, then stir in the cumin and the reserved bacon. Stir in the orange juice and turn off the heat. (The recipe can be made up to an hour or so ahead to this point. Gently warm the dressing again before proceeding.)

3 Put the spinach in a bowl large enough to comfortably toss the salad quickly. Add the sweet potatoes and the warm dressing and toss to combine. Taste and adjust seasoning, and serve.

Hummus with Pita and Greens

Makes: **4 servings** Time: **About 25 minutes with cooked chickpeas**

This is more salad than sandwich. I make this open-faced, with the crunchy pita and spread nestled under a pile of greens. But you can easily deconstruct the dish and serve the pita (toasted or not) alongside for scooping up the hummus. Or if you have pocket pitas, smear the insides with the hummus and fill with the stuffed greens for a more portable lunch. (To make just hummus, follow Step 2.)

4 whole wheat pitas

2 cups drained cooked (page 139) or canned chickpeas, some liquid reserved if possible

½ cup tahini (with some of its oil), or more to taste

2 cloves garlic, peeled, or to taste

⅓ cup olive oil

1 tablespoon ground cumin or smoked paprika (*pimentón*), more or less, plus a sprinkling for garnish

Salt and freshly ground black pepper

Juice of 1 lemon, plus more as needed

6 cups lettuce or assorted salad greens, torn into pieces

Cucumber slices, tomato wedges, thinly sliced red onion, and/or black olives, for garnish

½ cup chopped fresh mint or parsley leaves, for garnish

1 To toast the pitas if you like, heat the oven to 450°F. Put them on a baking sheet and cook until just barely crunchy on both sides, about 15 minutes total.

2 Meanwhile make the hummus: Combine the chickpeas, tahini, garlic, and ¼ cup of the oil in a food processor with the spice and a sprinkling of salt and pepper. Use the reserved bean liquid (or water) as necessary to get machine going. Purée, then add about half of the lemon juice, along with more tahini or salt if desired.

3 When the pita has cooled smear a layer of hummus on each and put on plates. (You'll probably have some left over; the hummus will keep, refrigerated, for about a week. Eat it with raw vegetables or on bread.) Put the lettuce in a bowl, sprinkle with some salt, pepper, and a pinch of the spice you used and drizzle with the remaining olive oil and lemon juice. Toss well then pile on top of the pitas. Garnish and serve.

Hummus with Dried Tomatoes: Omit the tahini. Cover ½ cup (or more) dried tomatoes with boiling water and soak until softened, about 15 minutes. Drain (reserve the liquid for another use) and follow the recipe, substituting the tomatoes for the tahini.

Vegetarian Sandwich Ideas

Once you buy into the idea that a sandwich doesn't have to include meat, you can enjoy all sorts of good stuff, whether slipped between two pieces of bread, served open-faced, wrapped in a tortilla, or stuffed in a pita pocket. Any of these ideas for veggie sandwiches will work with the breads in this book: Almost No-Work Whole Grain Bread (page 156), Hybrid Quick Bread (page 154), or Easy Whole Grain Flatbread (page 224). Or try warmed whole wheat or corn tortillas, pitas with or without the pocket, whole grain lavash, or store-bought whole grain loaves.

Greek Salad Sandwich: Romaine lettuce, tomatoes, olives, cucumbers, and feta cheese in a pita pocket, drizzled with olive oil and a squeeze of lemon.

Mediterranean Wrap: A whole wheat tortilla filled with either Not Your Usual Ratatouille (page 206) or Salade Niçoise with Mustard

Vinaigrette (page 186); or a simple mixture of white beans, arugula, olive oil, and chopped black olives.

Tex-Mex Sandwich: Whole wheat sandwich bread smeared with lightly mashed black beans (page 139), and topped with chopped avocado, tomatoes, chiles, red onions, and pickles.

Caprese Sandwich: Whole grain flatbread or lavash filled with thinly sliced mozzarella, thick slices of tomato, basil leaves, and Olive Oil Drizzle (page 158).

Garden Sandwich: Multigrain sandwich bread spread with a smear of hummus (page 196) or tahini mixed with a little water, and filled with tomatoes, sprouts, radishes, cucumber, and minced chives.

Leftover Sandwich: Any roasted or grilled vegetables, chopped up and served in pita pockets or rolled into a lavash or tortilla wrap. Garnish as you like with lettuce, tomatoes, pickles, onions, or other condiments.

ALT: Avocado, tomato, and lettuce on toasted whole wheat bread; smear of mayo optional.

Mushroom Sandwich: Grilled or broiled portobello mushroom with roasted garlic on whole wheat bread or bun.

Nut-Wich: Lightly mash something delicious, smear it on toasted bread, then sprinkle chopped nuts on it. Some excellent combos: banana, honey, and almonds; avocado and peanuts; sweet potatoes and hazelnuts; apricots and almonds; peaches and pistachios; roasted beets and walnuts.

Bread Spread: Any vegetable or bean spread (page 222), with lettuce, tomatoes, and thinly sliced onions.

Stir-Fried Beans with Asparagu or Broccoli

Makes: **4 Servings** Time: **15 minutes**

A different kind of stir-fry: no soy sauce. I like asparagus or broccoli, but in the winter, halved Brussels sprouts are excellent, especially with big white lima beans or chickpeas. Serve with some brown rice or toss with noodles.

1 pound asparagus or broccoli, trimmed

2 tablespoons olive oil

Salt and freshly ground black pepper

3 scallions, chopped

1 tablespoon minced garlic

1 teaspoon ground cumin

1/2 cup chopped dried tomatoes

1/2 cup vegetable stock (page 150), white wine, or water, plus more as needed

1 cup edamame, fava, or lima beans, fresh or thawed frozen

1 Cut the asparagus into 2-inch pieces or break the broccoli into florets. Put the oil in a skillet over medium-high heat. When the oil is hot, add the asparagus or broccoli, sprinkle with salt and pepper, and cook, stirring frequently, until coated in oil and just beginning to soften, about a minute. Add the scallions and garlic, and cook, stirring occasionally, just another minute. Add the cumin and tomatoes and give a good stir, then add the remaining ingredients.

2 Cook, stirring occasionally, until the tomatoes plump up, the liquid is reduced a bit, and the vegetables and beans are crisp-tender, about 5 more minutes. If you prefer more liquid for tossing with pasta or rice, add a little more, but be careful not to overcook anything. Remove from heat, taste and adjust the seasoning, and serve.

Fast Mixed Vegetable Soup

Makes: **4 to 6 servings** Time: **About 30 minutes**

Making a flavorful soup is as simple and easy as simmering vegetables in stock or water; if you cook them in olive oil to begin, so much the better. To build on the main recipe, see the list that follows.

3 tablespoons olive oil, or more

1 medium onion, chopped

1 carrot, chopped

1 celery stalk, chopped

2 garlic cloves, smashed and chopped

Salt and freshly ground black pepper

6 cups vegetable stock (page 150) or water

2 medium tomatoes, peeled, seeded, and chopped (canned tomatoes are fine; use 3 and include the juice), or 1/4 cup tomato paste

4 to 6 cups quick-cooking vegetables, like green beans, cauliflower, broccoli, asparagus, corn, cooked or canned beans, radishes, zucchini or summer squash, or dark, leafy greens like kale or collards, roughly chopped

1/2 cup chopped fresh parsley leaves

1 Put 2 tablespoons of the oil into a large, deep pot over medium heat. When it's hot, add the onion, carrot, celery, and garlic. Sprinkle with salt and pepper and cook, stirring, until the onion softens, about 5 minutes.

2 Add the stock, tomato, and remaining vegetables; bring to a boil, then lower the heat so the mixture bubbles enthusiastically.

Cook, stirring every now and then, until all the vegetables are very tender, 10 to 15 minutes. Taste and adjust the seasoning, add the remaining tablespoon of olive oil, and serve.

How to Vary Fast Mixed Vegetable Soup:

Up the quantity of beans, and mash some or all before adding if you like.

Spoon pesto or tapanade on top.

Sprinkle with croutons or bread crumbs.

Pass grated Parmesan at the table.

Add cooked pasta or grains at the last minute; simmer just long enough to heat through.

Add fresh or frozen peas, limas, or edamame at the last minute before serving.

Add some reconstituted porcini mushrooms (with their liquid) along with the vegetables; or use sautéed (or raw, sliced) fresh mushrooms.

Change the herbs: try mint, basil, chives, or cilantro instead of the parsley; or use tablespoon of fresh chopped sage or thyme.

Add some spices: like Hot Curry Powder, Fragrant Curry Powder, Chili Powder (pages 143 to 145) or single spices like ground cumin or coriander seeds to the onion mixture just before adding the stock.

Before adding the vegetables, put some chopped root vegetables (like turnips, potato, sweet potato, or celery root) into the pot. Let them cook for about 10 minutes, stirring occasionally, before proceeding.

Puree half or all of the soup just before serving (an immersion blender is ideal for this).

When you sauté the vegetables in Step 1, add ½ cup chopped prosciutto or other ham.

Creamy Carrot Soup

Makes: **4 servings** Time: **45 minutes**

This soup is as good cold as it is hot, and its creaminess comes from vegetables, not dairy items, though you can certainly enrich this soup by stirring in a pat of butter or a splash of cream or coconut milk after pureeing.

In place of the carrots you might try fennel or celery; root vegetables like parsnips, celery root, or turnips; spinach, sorrel, or watercress; sweet potatoes or winter squash; any potatoes; peas (alone or with some romaine lettuce). If you want to add spices—either curry powder on page 144 or ground cumin seeds are good with carrots—stir them in just before adding the stock in Step 1.

3 tablespoons olive oil

1 small onion or a 2-inch piece of ginger, chopped

About 1½ pounds carrots, roughly chopped

1 large starchy potato, peeled and roughly chopped

Salt and freshly ground black pepper

6 cups vegetable stock (page 150) or water

¼ cup chopped parsley leaves for garnish

1 Put the oil in a large, deep saucepan or Dutch oven over medium heat. When the oil is hot, add the vegetables. Sprinkle with salt and pepper and cook, stirring occasionally, for about 15 minutes, until the carrots soften a bit. Add the stock and cook until the vegetables are very tender, 15 to 20 minutes.

2 Use an immersion blender to puree the soup in the pan. Or cool the mixture slightly (hot soup is dangerous), and pass it through a food mill or pour it into a blender. Puree until smooth, working in batches if necessary. (You can make the soup ahead to this point. Cover, refrigerate for up to 2 days, and reheat before proceeding.)

3 If you're serving the soup hot, gently reheat it, stirring frequently. If you're serving it cold, refrigerate, covered, for at least 2 hours. Either way, taste and adjust the seasoning and garnish just before serving.

Chunky and Creamy Carrot Soup: Use ginger. In Step 2, puree only half of the soup and return it to the pot along with some chopped shrimp or ham, or tofu cubes. Add a dried chile to the pot if you like and cook until the shrimp is cooked or the ham or tofu is heated through. Garnish with chopped cilantro, scallions, and peanuts.

Curried Lentil Soup with Potatoes

Makes: **4 servings** Time: **About 45 minutes, mostly unattended**

You can use any small beans you like in this soup, which has a variety of flavors and textures. Red lentils will cook even faster; others (whole peas, mung beans, or small navy beans) might take up to 15 minutes more.

I've offered a couple of different vegetable options, but feel free to add on. You can also use turnips, carrots, parsnips, or other root vegetables in place of some or all of the potatoes. Shredded cabbage and sliced okra are good choices. A bit of chicken is also nice: if you have some already cooked, chop or shred it and add it to the pot at the last minute. (To cook chicken along with the lentils, see the variation.)

2 tablespoons peanut or grape seed oil

1 medium onion, roughly chopped

1 tablespoon minced garlic

1 tablespoon minced peeled fresh ginger

Salt and freshly ground black pepper

3 tablespoons curry powder (to make your own, see Hot Curry Powder and Fragrant Curry Powder, page 144)

2 medium tomatoes, peeled and seeded if you like, and chopped (or use 4 canned tomatoes)

1 cup dried lentils, washed and picked over

1 quart vegetable stock (page 150) or water, plus more as needed

1 can coconut milk, or another 1½ cups of stock or water

2 medium russet or sweet potatoes, peeled and cut into chunks

1 small zucchini, roughly chopped; or 1 cup chopped green beans or carrots

½ cup chopped fresh cilantro or mint leaves

1 Put the oil in a deep skillet or medium saucepan over medium-high heat. When hot, add the onion and cook, stirring occasionally, until soft and translucent, about 3 minutes. Add the garlic and ginger and cook for another minute. Sprinkle with salt and pepper and stir in the curry powder. Cook, stirring frequently, until darkened and fragrant, another minute or two.

2 Stir in the tomato and lentils, then add the stock and coconut milk or water. Bring to a boil; partially cover, and turn the heat down to medium-low so that the soup bubbles gently.

3 Cook, stirring occasionally, until the lentils are just becoming tender; stir in the potatoes and more stock or water if needed. Cover again and cook for about 10 minutes, then stir in the remaining vegetables, adding a little more water if needed to keep everything brothy. Cover one more time and cook until the potatoes and vegetables are all tender, another 5 to 10 minutes more. Stir in the cilantro or mint, taste and adjust the seasoning, and serve.

Curried Lentils with Potatoes (Dal): Thicker, less soupy, and excellent over rice or to serve with Easy Whole Grain Flatbread (page 232): Omit the tomatoes and zucchini, green beans, or carrots. Reduce the stock to 2 cups. Proceed with the recipe, garnishing with chopped unsweetened coconut if you like.

Curried Lentil Stew with Chicken and Potatoes: Reduce the stock to 2 cups. Sprinkle 4 chicken thighs with salt and pepper. In Step 1, after you heat the oil, add the chicken to the pan and cook, turning once, until both sides are nicely browned. Remove the chicken from the pan and proceed with the recipe through cooking the lentils in Step 3. When you add the potatoes, return the chicken to the pot and finish the stew as directed.

Not Your Usual Ratatouille

Makes: **4 to 6 Servings** Time: **30 minutes**

I've got nothing against zucchini (obviously: it appears in several recipes in this book), but cauliflower makes a much more interesting contribution to ratatouille, because its crunchy chew complements the creaminess of the eggplant beautifully. If you prefer, replace the cauliflower with 2 or 3 small zucchinis or 1 large zucchini or summer squash. To add more substance here, stir in some white beans, cut pasta, roasted potatoes, or bits of chicken or feta cheese as the cooking nears its end.

1 medium or 2 small eggplants (about 8 ounces)

Salt

1/4 cup olive oil

Freshly ground black pepper

1 small head cauliflower, trimmed and cut into florets

1 small onion, chopped

1 tablespoon minced garlic

1 red bell pepper, cored and chopped

2 medium tomatoes, cored and chopped

1 tablespoon chopped fresh thyme

1/2 cup chopped basil leaves, for garnish

Good vinegar or freshly squeezed lemon juice, optional

1 Trim the eggplant and cut it into large cubes. If the eggplant is big, soft, or especially seedy, sprinkle the cubes with salt, put them in a colander, and let them sit for at least 30 minutes, preferably 60. (This will help improve their flavor, but isn't necessary if you don't have time.) Then rinse, drain, and pat dry.

2 Put 2 tablespoons of the oil in a large skillet over medium heat. When hot, add the eggplant, sprinkle with salt and pepper, and cook, stirring occasionally, until soft and golden, about 10 minutes. Remove from the pan and drain on paper towels.

3 Put the remaining 2 tablespoons oil in the pan and add the cauliflower. Cook, stirring occasionally, until it loses its crunch, about 4 minutes. Add the onion, garlic, and red pepper and cook and stir for another minute or two, until they're soft. Add the tomato and thyme and cook for another minute, until the tomato just starts to release its juice. Return the eggplant to the pan, along with basil leaves. Give a good stir, taste and adjust the seasoning, and serve hot or at room temperature, with vinegar or lemon. The ratatouille will keep for a couple of days, covered and refrigerated.

Impromptu Fried Rice

Makes: **4 servings** Time: **20 minutes, with cooked rice**

Fried rice is an ideal way to use leftovers. In fact, it doesn't work well with freshly made rice; the kernels have to be dry from spending some time in the fridge. This recipe specifies fresh vegetables, but use whatever you have and simply cook them long enough to just reheat. If you want some meat in there, as little as an eighth of a pound will make a difference—use ham, bacon, chicken, or whatever else you like, browned separately. Generally, the trick is to stir-fry the ingredients in stages, very quickly, so everything fries instead of steaming.

3 tablespoons peanut or vegetable oil

1 bunch scallions, cut into 1-inch pieces

1 red bell pepper, cored, seeded, and roughly chopped

½ pound green beans or asparagus, cut into 1-inch pieces

1 cup fresh or thawed frozen peas (optional)

1 cup soy or mung bean sprouts (optional)

1 tablespoon minced garlic, or to taste

1 tablespoon peeled and minced fresh ginger, or to taste

3 to 4 cups cooked brown rice (page 136)

2 eggs, lightly beaten (optional)

¼ cup sherry, white wine, stock, or water

2 tablespoons soy sauce

1 tablespoon dark sesame oil

Salt and freshly ground black pepper to taste

¼ cup chopped fresh cilantro leaves for garnish (optional)

1 Put 1 tablespoon of the oil in a large skillet over high heat. A minute later, add the scallions and bell pepper and cook, stirring occasionally, until they soften and begin to brown, about 3 minutes. Lower the heat if the mixture starts to brown too quickly. Transfer the vegetables to a bowl.

2 Add the green beans or asparagus and cook, again over high heat, stirring occasionally, until nicely browned and just tender, about 5 minutes. Add them to the bowl with the vegetables. If you're using the peas, drain them if necessary and add them, along with the bean sprouts (if you're using them), to the skillet; cook, shaking the skillet, for about a minute, or until steaming. Add them to the bowl, too.

3 Put the remaining oil in the skillet, followed by the garlic and ginger. Just a few seconds later, begin to add the rice, a little at a time, using a spatula to break up the clumps, and toss it with the oil. When all the rice is added, make a well in its center and break the eggs into it if you're using them; scramble them a bit with the spatula, then stir to incorporate them into the rice.

4 Return the vegetables to the pan and use the spatula to fold everything until combined. Add the wine and cook, still folding, for about a minute. Add the soy sauce and sesame oil, then taste and add salt and pepper if necessary. Turn off the heat, stir in the cilantro, and serve.

Quick Vegetable Fried Grains: Instead of the rice, use quinoa, cracked wheat, oat groats, or large-kernel grains like wheat berries or barley. Just make sure they're relatively dry before starting.

Other Ingredients to Try in Fried Rice

Substitute any of these for some or all of the ingredients in the recipe.

- Grated cabbage
- Chopped carrots
- Minced hot chiles
- Snow peas or snap peas
- Peeled shrimp or squid
- Bits of smoked sausage or ham, or cooked pork or chicken

Pan-Cooked Greens with Tofu and Garlic

Makes: **4 Servings** Time: **30 minutes (more if freezing the tofu beforehand)**

With thick-stemmed greens, you get two vegetables for the price of one: the leaves cook up tender and silky, while the center ribs remain crunchy. You have to take the time to separate them, but it really does make a difference. And do try freezing the tofu overnight (or up to a couple of months); when it is thawed, its texture is dense and meaty.

You can eat this dish hot or at room temperature like a warm salad, with brown rice (page 136). Or roll everything into a big whole wheat tortilla.

- 1 pound firm tofu
- 1½ pounds of kale, Napa cabbage, bok choy, or other thick-stemmed Asian greens
- 3 tablespoons peanut or vegetable oil, plus more if needed
- 3 tablespoons nam pla (Thai or Vietnamese fish sauce), or soy sauce
- 1 small chile, stemmed, seeded if you like, and minced, or crushed red chile flakes to taste
- 1 tablespoon sugar
- 2 tablespoons lime juice or rice vinegar, or to taste
- 2 or 3 cloves garlic, cut into slivers
- Salt and freshly ground black pepper

1 If time allows, freeze the tofu and thaw it out before starting. Either way, squeeze the block of tofu between your palms (over a bowl or the sink). Be firm but not too enthusiastic: you want to push out some excess water without smashing the tofu. Cut in half lengthwise, then slice into thin pieces and set aside. For the greens, separate the stems from the leaves; cut the stems into 1-inch sections and roughly chop the leaves.

2 Put a tablespoon of the oil in a large skillet over high heat, add about half the vegetable stems, and cook, stirring frequently until they're browned and slightly tender, 3 to 5 minutes. Remove to a bowl with a slotted spoon and repeat with remaining stems. Remove and repeat with leaves. Remove. While greens cook, combine nam pla, chile, sugar, and lime juice or vinegar in a small bowl.

3 Add a little more oil to skillet if necessary, followed by the tofu slices; work in batches if the pan is too crowded (these won't take long to cook). Cook, flipping them once with a spatula, until browned on both sides, less than 5 minutes. During the last minute or so of cooking, add the garlic, return the greens to the pan, and stir. Turn off the heat, drizzle with the dressing, and toss again. Taste and add salt if necessary and lots of black pepper. Serve hot or at room temperature.

Stir-Fried Greens with Cashews: Instead of the tofu, use 1 cup whole cashews. Proceed with the recipe.

Microwaved Greens with Tofu: Cut the tofu into cubes instead of slices. Start by mixing the dressing ingredients together as described in Step 2, but in a big bowl. First put the stems in a microwave-proof bowl or plate fitted with a lid or a piece of vented plastic. Cook on high for about 3 minutes (more or less, depending on your machine), until the stems are just becoming tender. Transfer to the bowl with the dressing and toss. Repeat the process with the leaves; this will take only a couple of minutes. Then cook the tofu in the microwave until just steaming, another minute or two. Toss everything together in the bowl and taste. Season and serve as described in Step 3.

Noodles with Mushrooms

Makes: **4 servings** Time: **30 minutes**

Mushrooms have an earthy flavor and a chewy texture that people often describe as "meaty." And they are, especially if you add dried mushrooms into the mix. This is a wonderfully satisfying vegetarian pasta dish, whether you go for the Italian profile of the main recipe, or the Asian-style variation.

- ¼ to ½ cup dried porcini mushrooms (optional)
- 1 cup hot water (optional)
- Salt
- 1 pound fresh mushrooms (shiitakes are nice here; remove the stems and save them for another use)
- ¼ cup plus 1 tablespoon olive oil
- Freshly ground black pepper
- 2 tablespoons minced shallot or 1 tablespoon minced garlic
- 1 pound dried pasta, preferably whole wheat
- About ½ cup chopped fresh parsley leaves, plus more for garnish

1 Bring a large pot of water to a boil and salt it. If you're using the porcini, put them in a small bowl, cover with the hot water, and set aside to soak for about 15 minutes. (If you're not using them, skip to Step 2.)

2 Rinse the fresh mushrooms and trim off any hard, tough spots; cut them into small chunks or slices. If you're using the porcini, lift them out of the soaking water; save the water, undisturbed, so that the sediment settles on the bottom of the bowl.

3 Put ¼ cup of the oil in a medium to large skillet over medium heat. When the oil is hot, add all of the mushrooms and sprinkle with salt and pepper. Raise the heat to medium-high and cook, stirring occasionally, until the mushrooms begin to brown, at least 10 minutes. Add the shallot or garlic and stir until the mushrooms are tender, another minute or two. Turn off the heat.

4 Cook the pasta until tender but not mushy, from 6 to 8 minutes. When it's almost done, add about ½ cup of the pasta cooking water to the mushrooms (or use the porcini soaking liquid, being careful to leave the sediment in the bowl), turn the heat to low, and reheat gently. Drain the pasta, reserving a little more of the cooking water. Toss the pasta and the mushrooms together with the remaining tablespoon of olive oil; add a little of the pasta cooking water (or porcini liquid) if the dish seems dry. Taste and adjust the seasoning. Stir in the parsley and serve garnished with more parsley.

Other Dishes in the Book You Can Eat for Lunch

Breakfast Burritos (page 176)

More-Vegetable-Than-Egg Frittata (page 170)

Breakfast Couscous (page 166)

Savory Bread Pudding (page 175)

Easy *Socca* or *Farinata* (page 225)

Easy Whole Grain Pizza (page 225)

Vegetable Pancakes (page 230)

Stir-Fried Vegetables with Shellfish or Meat (page 238)

Orchiette with Broccoli Rabe, My Style (page 251)

Bulgur Pilaf with Vermicelli, and Meat or Cauliflower (page 258)

Bean and Vegetable Chili (page 260)

Asian-Style Noodles with Mushrooms: If using dried mushrooms, use dried shiitakes. For the fresh mushrooms, use either shiitakes or regular buttons. Substitute 3 tablespoons peanut oil and 1 tablespoon sesame oil for the olive oil; and omit the garlic or shallot. Use soba noodles instead of pasta. Proceed with the recipe, reducing the noodle cooking time to about 5 minutes. If you like, garnish with a drizzle of soy sauce, sliced scallions, and sesame seeds.

Snacks and Appetizers

You probably already snack, but when you eat like food matters, snacks take on a different meaning, because you can eat them without guilt. This means you don't have to worry quite so much about getting completely full at meals, as long as you remember to have snack food available.

For me, the basic snack is an apple (and what I've found is that whenever I do keep apples around, I actually *do* eat one—or more—a day), a banana, or any other piece of fruit; a handful (or more) of nuts or seeds; some trail mix—in short, something I can grab on the way out the door. Most of these recipes are just a little more formal, and most of them can serve not only as snacks but as appetizers and even first courses.

There is a wide range of food here. At the most basic, there is popcorn and homemade Pita Triangles or Tortilla Chips. Just a little more challenging are bean and vegetables purees, which have become standbys in my kitchen (those pita or tortilla chips accompany them perfectly). *Pinzomonio* (page 220) is definitely a step up from crudités. As for handheld snacks, you'll feel no guilt about sending kids out the door with Fruit and Cereal Bites (page 232)—a kind of homemade power bar without the unnecessary ingredients (you'll eat it, too). And no one will be able to resist the (baked) Root Vegetable Chips (page 226).

Warm Nuts and Fruit

Makes: **4 to 6 servings** Time: **15 minutes**

Here I convert good old raisins and peanuts (GORP) into a more sophisticated snack, or even an appetizer, by toasting the nuts in the oven and tossing them—still hot—with dried fruit. My favorite combinations lean on almonds, cashews, and pistachios with a sprinkling of whatever else is on hand, including peanuts, coconut, sunflower seeds, or pumpkin seeds. For fruit, think of raisins, banana chips, dates, dried cranberries, apricots—you name it. (Or skip the fruit entirely and use twice as many nuts.)

You may want to double or even triple this recipe: it goes fast and keeps well. And you can always speed things up by putting the oiled nuts in a bowl and zapping them in the microwave for several minutes, stirring every minute or 2 before adding the fruit and seasonings.

2 cups (about 1 pound) mixed unsalted shelled nuts

2 tablespoons olive, peanut, or nut oil

1 cup mixed dried fruit

Salt to taste

1 Heat the oven to 450°F. Put the nuts on a baking sheet, drizzle with oil, and toss well until evenly coated. Roast, shaking the pan occasionally, until lightly browned, about 10 minutes.

2 Put the warm nuts in a bowl with the dried fruit, sprinkle with salt, and toss. Cool briefly and serve.

Herbed or Spiced Warm Nuts and Fruit: When you toss the fruits and nuts together, add any of the following along with the salt: 1 tablespoon minced fresh rosemary, 1 teaspoon ground cumin, a pinch of cayenne pepper, ½ teaspoon ground cardamom or cinnamon, or ¼ teaspoon ground nutmeg or cloves.

Salty-Sweet Warm Nuts and Fruit: Add 1 tablespoon coarse sea salt and 1 tablespoon brown sugar to the butter.

Chili Nuts: Omit the fruit and double the quantity of nuts; toss the toasted nuts in a bowl with the salt and 2 teaspoons chili powder (to make your own, see page 143).

Pinzomonio: Crudités You Actually Want to Eat

Makes: **8 servings** Time: **30 to 60 minutes, depending on how involved you want to get**

Done right, crudités should bear no resemblance to the pathetic dried-up celery sticks and sour cream soup-mix dip you see at office parties. Instead, this dipping sauce and its variation are based on two Italian appetizers: *pinzomonio* and *bagna cauda*. One key is to use the best olive oil you can lay your hands on; another is to serve a wide variety of the very freshest raw or just-cooked vegetables.

You can use whole cherry tomatoes, jicama or carrot sticks, sliced celery and fennel, radishes, endive spears, and sugar snap peas. Vegetables that are strong-flavored, too tough when raw, or simply not very enjoyable raw—asparagus spears, string beans, small potatoes, and so on—should be lightly steamed or boiled, pulled from the water while still crisp, and shocked in a bowl of ice water (page 133). Then there are the in-betweens: broccoli, cauliflower, beets, and other root vegetables. Thinly sliced (or when small) these can be delicious raw, though some people prefer them slightly cooked. Once you have trimmed and cooked them as needed, cut the vegetables into manageable pieces.

Prepare as few or as many vegetables as time allows, and store raw vegetables in ice water to keep them crisp, and keep the barely cooked ones in an airtight container; both will hold well for a day or so. Drain and dry the vegetables before serving, and let them come to room temperature. If you're making *bagna cauda*, prepare the oil an hour or two in advance and simply reheat it before serving; use a fondue pot if you've got one, but an earthenware dish is fine, too.

3 to 4 pounds assorted crudités (see the headnote)

1 cup olive oil

Salt and freshly ground black pepper

1 Prepare all the vegetables as described in the headnote and store for later or put into serving pieces (small bowls, glassware, and platters all work).

2 Mix the oil with a large pinch of salt and put it in one or two bowls. Serve the vegetables with the oil as a dip.

Bagna Cauda: This is a little more complicated. Combine in a saucepan (or a fondue pot) 4 ounces anchovy fillets, packed in olive oil; 1 tablespoon minced garlic; and 1 tablespoon fresh chopped rosemary or savory (or 2 teaspoons dried). Turn the heat as low as possible. Cook, stirring constantly, until the anchovies break up, about 10 minutes (do not let the garlic brown). Add lots of freshly ground pepper and transfer the dip to an earthenware dish or set the fondue pot over its burner. Taste and add a bit of salt if necessary; you may not need any. (Keep warm or set aside for up to a couple of hours, then reheat just before serving.) Serve the warm dip alongside the vegetables.

Vegetable Spread

Makes: **4 servings** Time: **About 40 minutes, including cooking the vegetables**

Baba ghanoush, the classic Middle Eastern eggplant dip, is the model for this dish. However, I've turned the procedure into a master recipe that applies to nearly any vegetable: zucchini, butternut or other winter squash, cauliflower, broccoli, and most root vegetables like carrots, celery root, and parsnips. Greens, shell peas, and cooked beans work well, too. (For hummus, see page 202; but if you want a straight bean dip, substitute about 3 cups of cooked beans for the vegetables here.)

Choose flavorings from the list that follows, and the possibilities are almost endless. Here are some combos to get you started: ginger is killer with butternut squash, dill adds a lovely complexity to zucchini, mint brightens fava beans, and mustard seeds lend sharpness to cauliflower.

It doesn't matter whether you steam, boil, sauté, grill, or roast the vegetables first, though grilling and roasting concentrate flavors and make the spread more complex; just make sure everything is quite tender. Use the finished spread or dip as an appetizer or for sandwiches, or gently reheat and serve over rice or toss with pasta.

About 2 pounds any vegetables, trimmed and cooked until tender by any method (page 132)

3 tablespoons olive oil, plus more as needed

Salt and freshly ground black pepper

1 Make sure the vegetables are relatively dry before starting. If you need to drain them, reserve the cooking liquid. To puree the vegetables, put them in a blender or food processor with the olive oil and as much of the cooking liquid (or water or more olive oil) as you need to get the machine going; or run the vegetables through a food mill. (In many cases, you can simply mash the vegetables with a large fork or potato masher, adding the olive oil and cooking liquid as needed to reach the consistency you want.)

2 Taste, then sprinkle with salt and pepper and taste again. Serve warm, cold, or at room temperature. (The mixture will keep in the fridge for several days.)

Some Ways to Flavor Vegetable Spread

Add up to ½ cup of fresh parsley, mint, dill, cilantro, basil, or other mild herb leaves before pureeing.

Add up to a tablespoon of fresh rosemary, oregano, or thyme leaves before pureeing.

Squeeze some citrus juice—lemon, lime, or orange—into the puree.

Include a few coins of peeled fresh ginger or a garlic clove or two with the vegetables as they puree.

Puree the vegetable mixture with fresh or reconstituted dried chiles to taste, or add a pinch of cayenne or red pepper flakes.

When you add salt, add a pinch of ground ginger, cinnamon, cardamom, mustard seeds, or nutmeg.

Add either Chili Powder (page 143) or Med Mix (page 144) along with the olive oil.

Instead of the olive oil, use peanut oil or coconut milk, and season with either Hot or Fragrant Curry Powder (page 144).

Instead of the olive oil, use a combination of sesame oil and peanut oil, and season with either Five-Spice Powder or Sesame Shake (page 145).

Puree the vegetable mixture with a spoonful or two of yogurt, sour cream, cream, or crème fraîche.

Easy Whole Grain Flatbread

Makes: **4 to 6 appetizer servings** Time: **About 45 minutes, largely unattended (longer for resting, if time allows)**

The simplest bread is nothing more than water and flour. Heat some olive oil in a pan—you can add other flavorings, too—and this basic formula becomes a quick flatbread that's ready in the time it takes to cook dinner. The idea comes from the recipe for *socca* (also called *farinata*), the Mediterranean "pizza" made from chickpea flour (see the variation below). Chickpea flour and buckwheat flour are certainly options for the main recipe, too, but whole wheat flour and cornmeal are far more common and equally delicious.

A couple of technical details. The resting time for the batter is optional, but it results in a more complex flavor and a creamier, less gritty texture. If you're in a hurry, though, just let the batter sit while the oven heats. It's still awesome. And though a round pizza pan with a lip is ideal, a 10- or 12-inch skillet also works well; the bread in the smaller pan will need less oil, will be a slightly bit thicker, and will take another 5 or 10 minutes to bake.

You can bake the bread up to several hours in advance; warm it a little if you like—or not.

1 cup whole wheat flour or cornmeal, or chickpea flour (also called *besan*; sold in Middle Eastern, Indian, and health food stores)

1 teaspoon salt

4 tablespoons olive oil (see the headnote)

½ large onion, thinly sliced (optional)

1 tablespoon fresh rosemary leaves (optional)

1 Put the flour into a bowl; add salt; then slowly add 1½ cups water, whisking to eliminate lumps. Cover with a towel, and let sit while oven heats, or as long as 12 hours. The batter should be about the consistency of thin pancake batter.

2 When ready to bake, heat the oven to 450°F. Put the oil in a 12-inch rimmed pizza pan or skillet (along with the onion and rosemary if you're using them) and put in the heated oven. Wait a couple of minutes for the oil to get hot, but not smoking; the oil is ready when you just start to smell it. Carefully remove the pan (give the onions a stir); then pour in the batter, and return the skillet to the oven. Bake 30 to 40 minutes, or until the flatbread is well browned, firm, and crisp around the edges. (It will release easily from the pan when it's done.) Let it rest for a couple minutes before cutting it into wedges or squares.

Easy Whole Grain Pizza: When the bread is done, top as you would pizza, using a relatively light hand. Smear a thin layer of tomato sauce on first if you like, then add a sprinkling or crumble of cheese and thinly sliced vegetables, cooked meat, olives, onions—whatever. Turn on the broiler and put the pan under the heat until the ingredients are hot and bubbly. Let rest as above, then cut and serve.

Easy *Socca* or *Farinata*: Crisp on the bottom, custardy on top; chickpea flour is authentic, but whole wheat flour produces lovely results. You'll need a deep 12-inch pan or skillet. Increase the water to 3 cups and add up to another 2 tablespoons of oil if you like. Bake as above, but longer—closer to an hour. To get a crisper top, set under the broiler for a couple of minutes after the bottom is nicely browned. Let cool a bit in the pan, then slide a narrow spatula under the bottom to remove it. Cut into wedges and serve.

Root Vegetable Chips

Makes: **4 servings** Time: **30 minutes**

It's hard to beat potato chips, but since I've spent the first half of the book trying to talk you out of junk food, it seems only fair to offer you a great alternative. Most root vegetables will work here; beets are wonderful, but try potatoes, carrots, parsnips, rutabagas, kohlrabi, sweet potatoes, yams, and turnips—all are terrific. Cooking times will vary a bit depending on the vegetable and thickness; just turn them or pull them off the sheet pans whenever they're ready.

Dusted simply with salt and pepper or one of the Six Seasoning Blends You Can't Live Without (page 143), these crisps are far superior to anything that comes from a store or a vending machine.

1 pound beets or other root vegetables (see above), trimmed and peeled

3 to 4 tablespoons olive, peanut, or vegetable oil

Salt and freshly ground black pepper

1 Preheat the oven to 400°F. Lightly grease a couple of baking sheets or line them with parchment if you like. (For extra-crisp chips, heat the pans in the oven while you prepare the vegetables; then prepare the pans.)

2 Cut the beets in half and then crosswise into thin slices (1/8 inch or so). You can use a mandoline for this; just don't set it too thin, or the slices will stick to the pan. (If the beets are small, simply cut them crosswise.) Gently toss them in the oil and spread the slices out on the baking sheets; it's OK if they're close, but don't let them overlap.

3 Roast the beet slices until they're beginning to brown on the bottom, 10 to 12 minutes. Flip them over and sprinkle with salt and pepper or other seasonings. Keep roasting until they're well browned, another 10 minutes or so. Serve immediately.

Crisp Nori Ribbons

Makes: **4 servings** Time: **About 30 minutes, mostly unattended**

Nori, the thin sheets of dried seaweed most commonly used to wrap sushi, also makes terrific chips. When gently roasted, the sheets darken and become light, crisp, and a little curly, perfect as a light appetizer, with a bowl of edamame, as a snack, or as a crunchy garnish for many Asian dishes. Season with a good sprinkling of sea salt, a dash of Five-Spice Powder or Sesame Shake (page 145), plain sesame seeds, or a light brushing of sesame oil.

6 nori sheets

Sea salt or other seasonings, to taste

1 Heat the oven to 250°F. Brush or spray one sheet of the nori very lightly with a small amount of water and sprinkle with sea salt and any other seasonings. Press a second sheet on top of the first like a sandwich; the sheets won't bind perfectly, but they should stay together well enough to get you through Step 2.

2 Use a pizza cutter or sharp knife to cut the layered nori into 1- by 3-inch ribbons (or any desired size). Transfer to a cookie sheet in a single layer. Bake the ribbons for about 20 minutes; they will curl, crisp, and darken as they cook. When the chips are done, they become superlight and break easily, so remove the pan from the oven carefully and slide them off the cookie sheet onto racks (they cool quickly), then serve immediately or store in a sturdy covered container for up to a day or so.

Big Beans with *Skordalia*

Makes: **8 servings** Time: **15 minutes, with precooked beans**

Gigantes and other big beans—lima, edamame, fava, and whatever else you can get—make great finger food. Fresh beans are preferable, but hard to find (so when you do, grab them). Home-cooked dried beans are also fine. Whichever you use, cook them only until just tender, with skins intact.

Serve the beans with toothpicks for spearing and dipping in *skordalia,* an eggless mayonnaise substitute from Greece. This version is emulsified with bread and nuts, and works beautifully with any crudité (page 220), though I find the beans a nice match with the nutty sauce. To make this recipe an integrated dish, simply toss the beans with the sauce.

- 4 cups cooked large beans (page 139)
- 1 thick slice day-old whole grain bread
- About 1 cup vegetable stock (page 150) or water
- 2 tablespoons olive oil, plus more as needed
- 1 cup pine nuts, walnuts, blanched almonds, or hazelnuts
- 2 cloves garlic, peeled, or to taste
- 1/4 teaspoon cayenne, or to taste
- 1 tablespoon freshly squeezed lemon juice, or to taste
- Salt and freshly ground black pepper

1 Put the bread in a bowl and saturate it with the liquid. Wait a couple minutes, squeeze most of the moisture out of the bread into the bowl, then put it in a food processor with the oil, nuts, garlic, and cayenne. Process the mixture until smooth; then, with the machine running, pour in enough of the remaining liquid to form a creamy sauce.

2 Add the lemon juice and some salt and pepper and serve immediately as a dipping sauce for the beans or other vegetables. Stored in an airtight container and refrigerated, the sauce will keep for up to 2 days.

Vegetable Pancakes

Makes: **4 servings** Time: **At least 30 minutes**

A surefire way to get anyone to eat any vegetable, these crisp babies are delicious as a side dish, alone as an appetizer, or served on a bed of Nicely Dressed Salad Greens (page 135) as lunch.

Root vegetables are most common, but you can use whatever looks good to you, alone or in combination: zucchini, yellow squash, winter squash, corn, or chopped scallions; even spinach or chard is good (just cook it, squeeze it dry, and chop it first). And consider tossing in a tablespoon of fresh herbs or spices. Sweet potato and corn benefit from a bit of cilantro, zucchini comes to life with dill, and ginger or cardamom will warm up winter squash beautifully. Serve with Olive Oil Drizzle (page 158), a sprinkling of Parmesan or chopped nuts, or any salsa.

About 1½ pounds grated vegetables, peeled first if necessary (3 cups packed), and squeezed dry

½ small onion, grated; or 4 scallions

1 egg or 2 egg whites, lightly beaten

¼ cup white or whole wheat flour, more or less

Salt and freshly ground black pepper

Olive or vegetable oil or butter for greasing the pan

1 Heat the oven to 275°F. Grate the vegetable or vegetables by hand or with the grating disk of a food processor. Mix together the vegetables, onion, egg, and ¼ cup of the flour. Sprinkle with salt and pepper. Add a little more flour if the mixture isn't holding together.

2 Put a little butter or oil in a large skillet or griddle over medium-high heat. When the butter is melted or the oil is hot, drop in spoonfuls of the batter, using a fork to spread the vegetables into an even layer, then press down a bit. Work in batches to prevent overcrowding. (Transfer finished pancakes to the oven until all are finished.) Cook, turning once, until nicely browned on both sides, about 5 minutes. Serve hot or at room temperature.

Fruit and Cereal Bites

Makes: **About 3 dozen bite-size balls** Time: **About an hour, largely unattended**

You can make this nutritious alternative to so-called energy bars with any dried fruit and practically any cereal, and it takes no time at all. To gild the lily, roll the finished bites in shredded unsweetened coconut, finely ground nuts, or cocoa.

- 1½ cups dried fruit
- 2 tablespoons vegetable oil
- 2 tablespoons honey (optional)
- Fruit juice or water, as needed
- 1 cup ready-to-eat breakfast cereal, like granola (page 168)
- Crumbled shredded wheat, or any whole grain flakes or "nuts"

1 Put the dried fruit, oil, and honey if you are using it in a food processor and puree until smooth, adding fruit juice a little at a time to keep the machine running. You'll need to stop once or twice to scrape down the sides of the bowl. (Add small amounts of water or fruit juice if the fruit is dried out and is not processing.) Fold in the cereal until evenly distributed.

2 Take a heaping tablespoon of the mixture and roll it into a ball. Then if you like, roll the ball around in any extra ingredients; see the headnote. Put the balls between layers of waxed paper in a tightly covered container and refrigerate until set, about 45 minutes. Eat immediately, or store in the fridge for up to several days. You can also wrap the balls individually in wax paper, like candies.

Fruit and Cereal Bars: Line an 8-inch by 8-inch pan with foil. Follow the recipe through Step 2. Spread the mixture in the pan, pushing it into the corners and evening the top. Refrigerate to set as above. If you like, dust the top with any of the other ingredients as described in the headnote. Then cut into squares.

Other Dishes in the Book You Can Eat for Snacks or Appetizers

Salsa, Any Style (page 146) with Baked Pita Triangles, Tortilla Chips, or Croutons (page 234)

Roasted Red Peppers (page 152)

Hybrid Quick Bread (page 154)

Almost No-Work Whole Grain Bread (page 156)

Fruit Smoothies (page 163)

Anything Goes Granola (page 168)

Any of the salads on pages 180 to 195

Any of the soups on pages 200 to 205

Not Your Usual Ratatouille (page 206) on whole grain crackers or toasted bread

Grilled or Broiled Kebabs (page 246)

Roasted Herb-Stuffed Vegetables (page 272)

Meat-and-Grain Loaves, Burgers, and Balls (page 278)

Savory Vegetable and Grain Torta (page 280)

Baked Pita Triangles, Tortilla Chips, or Croutons

Makes: **4 servings** Time: **About 30 minutes**

The same technique works perfectly for pita, tortillas, and croutons—and all are infinitely preferable to what comes in a package. Skip the oil if you like, though it not only adds flavor but also helps the seasonings stick. I often dust these with chili or curry powder (pages 143 and 144) and plenty of freshly ground black pepper, but use whatever spices you like.

If you're using pocket pita bread, split each piece into two circles first; these will be thinner and take less time to cook—but watch them carefully, as they burn easily.

2 or 3 whole wheat pitas, 8 small corn tortillas, or 1 whole grain baguette

Olive oil as needed (optional)

Salt

Freshly ground black pepper (optional)

1 Heat the oven to 450°F. Cut the pita or tortillas into 6 or 8 triangles, or cut the bread into ½-inch slices or cubes. Put the pieces on 1 or 2 rimmed baking sheets; it's OK if they're close but not overlapping. Bake, undisturbed, until the bread begins to turn golden, about 15 minutes. Drizzle, brush, or spray with olive oil if you like.

2 Turn the slices over, add more olive oil if you're using it, and continue baking until they're the desired color, anywhere from 5 to 15 minutes. Sprinkle with salt and pepper—or other seasoning—as you like. Serve immediately with hummus or salsa, or alongside soup or salad (or tossed *in* salad). Or store in an airtight container; these will stay crunchy for up to a week.

Brown-Bag Popcorn

Makes: **2 to 4 servings** Time: **5 minutes or less**

Popcorn is the perfect snack: It's a whole grain that costs next to nothing, can be made in minutes, and flavored any number of different ways. Buttered popcorn is probably still the best, but in the middle of the afternoon when I need something easy to munch on, this microwave version (which can also be made without oil or butter) does the trick—and you avoid the nasty additives present in packaged microwave corn. (This method also works in a deep glass casserole with a lid.)

I usually get lazy and just have this popcorn with salt, but you can jazz it up by adding superfine sugar for a salty-sweet mix, a handful of chopped fresh herbs, or one of the Six Seasoning Blends You Can't Live Without, on page 143.

¼ cup fresh, good-quality popcorn

¼ teaspoon salt, or less if you want

2 teaspoons peanut or vegetable oil

In a brown paper lunch bag, combine the popping corn and salt and oil; then fold the top of the bag over a couple times. Put it in the microwave on high for about 2 or 3 minutes, or until there are 4 or 5 seconds between pops. Open the bag carefully, because steam will have built up. Toss with your favorite seasonings or serve as is.

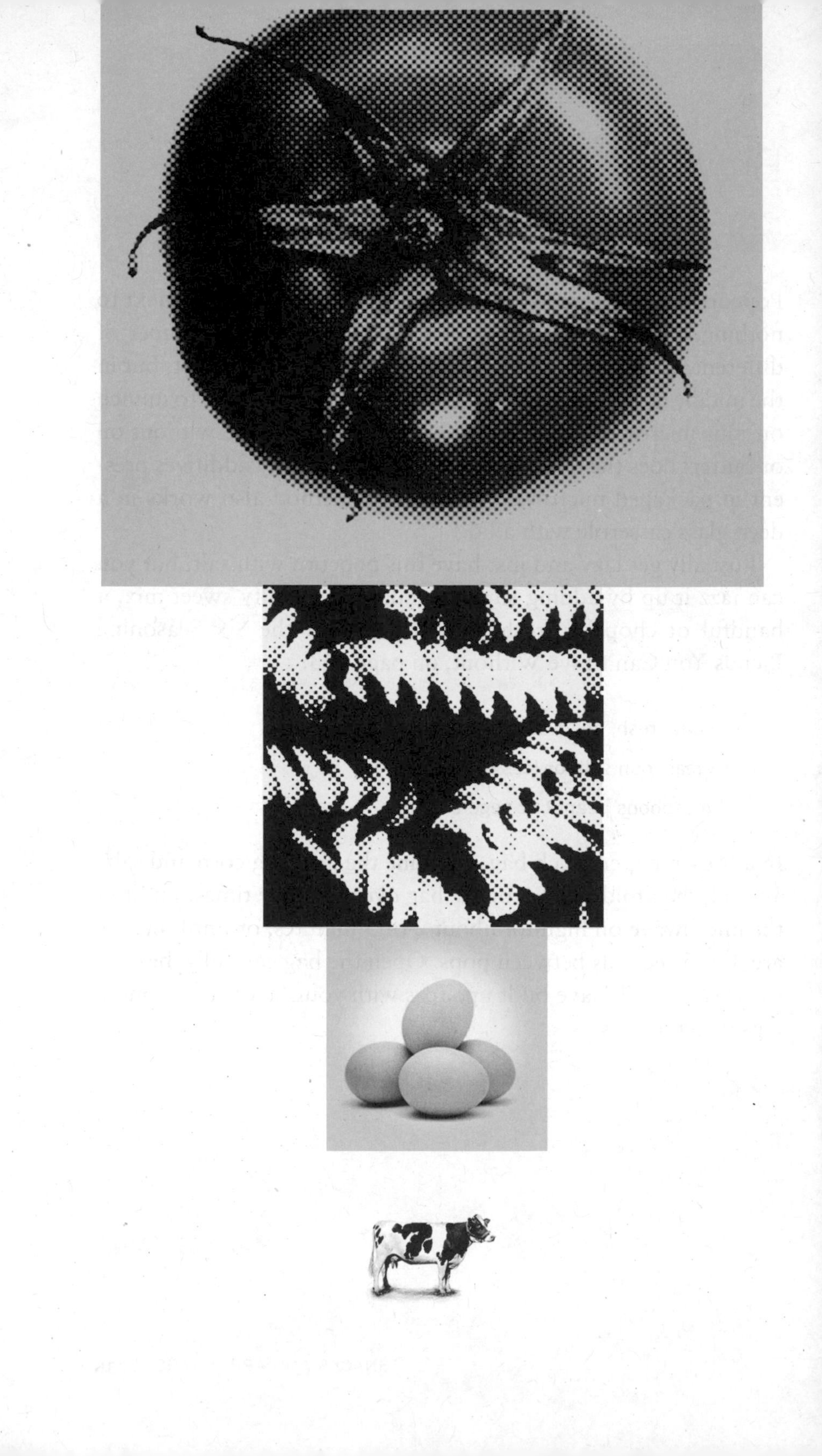

Dinner

Most people do most of their cooking at dinner, and it's at dinner we most often eat straight cuts of meat. These, then, are the Food Matters recipes that are most obviously different from the standard (including my own) recipes, in which the proportion of animal to plant foods is literally turned upside down. Vegetables are always abundant and almost always in the forefront. Meat, poultry, and fish are often optional.

But it's also at dinner where you are most likely to eat most of your calories, and in keeping with that, these are the heartiest recipes in the book; it's not like you're going to starve here. If I do say so myself, these recipes are fantastic.

The basic stir-fry is a less-meat classic; there's an assortment of vegetable-first recipes cooked by different methods, many of which include a bit of meat for flavoring. There are more standard preparations, like kebabs (with a large vegetable component), pasta, and chili. Then there are redefined, modernized international standards, like cassoulet, paella, chicken "not" pie (no crust!), bouillabaisse, and eggplant and chicken parmesan.

And most of these recipes are extremely adaptable: You can take advantage of whatever is in season, and in every case I've given suggestions for substituting, so each recipe can be taken in a number of different directions.

Stir-Fried Vegetables with Shellfish or Meat

Makes: **4 servings** Time: **20 to 30 minutes**

Super-simple and amazingly versatile, this works with any combination of vegetables and virtually any protein. Cutting the vegetables into bite-size pieces avoids the extra step of parboiling, so everything happens in one pan. Just remember that soft, pliable vegetables like greens are going to cook in an instant, while firm, hard vegetables are going to take a few minutes to become tender. Poke and taste as you work until you get the hang of judging doneness visually, and you'll be fine.

For vegetables, try bell pepper, cabbage, bok choy, fennel, spinach, snowpeas or snap peas, asparagus, green beans, edamame or other fresh or frozen shell beans, mushrooms, carrots, broccoli, or cauliflower. I love this with shrimp, but try scallops, squid, lobster, or crab; thinly sliced boneless beef sirloin, or pork or lamb shoulder; cut-up boneless chicken breast or thighs; or firm tofu cubes.

For a change of pace, try tossing this stir-fry with hot soba or other noodles.

1 recipe brown rice (page 136)

About 1/4 cup peanut or vegetable oil

2 tablespoons minced garlic

1 tablespoon grated or minced fresh ginger

4 cups assorted vegetables, trimmed, peeled, and seeded as needed and cut into bite-size pieces

1 bunch scallions, sliced on the diagonal

1/2 pound shellfish, meat, or tofu

2 tablespoons soy sauce

Juice of 1/2 lime (optional)

Salt and freshly ground black pepper

½ cup any vegetable, shrimp, or fish stock (pages 150–51), white wine, sake, or water

Chopped cilantro, mint, or Thai basil (optional)

1 Start the rice. After preparing the vegetables, separate them according to what will cook fastest and what will take longer. If using only one vegetable, divide it into two batches.

2 Put a large, deep skillet over high heat. Add a tablespoon of the oil, swirl, and then add half the garlic and ginger. Cook for 15 seconds, stirring, then add one of the vegetables. Cook and stir until almost, but not quite done as you want it, from a minute or two for spinach to as much as 5 minutes for root vegetables. Transfer the vegetables to a bowl or platter and repeat with the remaining vegetables, adding more oil to the pan as needed to prevent sticking. Just before you take the last batch of vegetables out, add the scallions and give a good stir.

3 Turn the heat down to medium and add another tablespoon of oil to the pan, then the remaining garlic and ginger. Stir; then

Some Simple Additions to Any Stir-Fry

Add either chopped fresh chile or red chile flakes to taste, along with the garlic.

Add 2 tablespoons whole or chopped peanuts or cashews when you return the vegetables to the pan.

Add 1 teaspoon ground spices, especially Sichuan peppercorns, with the garlic.

Use nam pla (Thai fish sauce) in place of soy sauce.

Add about 1 tablespoon ground bean paste (you can buy it at Chinese markets) during the last minute of cooking.

Finish the dish with a tablespoon of toasted sesame seeds, simply sprinkled on top.

add the shellfish, meat, or tofu. Raise the heat to high, stir once, then let it sit for 1 minute before stirring again. Cook, stirring occasionally, until the protein has lost its pink color, about 3 minutes for shrimp and scallops and about 5 minutes for chicken and other meats.

4 Return the vegetables to the pan and toss once or twice. Add the soy sauce and lime juice if you're using it; toss again. Sprinkle with salt and pepper, then add the liquid. Raise the heat to high and cook, stirring and scraping the bottom of the pan, until the liquid is reduced slightly and you've scraped up all the bits on the bottom of the pan. Garnish with basil, cilantro, or mint and serve.

Roasted Vegetables with or without Fish or Meat

Makes: **4 servings** Time: **About 1 hour**

A spectacular way to take advantage of vegetables in season, since the technique remains the same whether you use root vegetables or anything else. Vary it depending on what you've got, but remember that cooking times will be shorter for more tender vegetables and longer for sturdier ones. This recipe makes a wonderful bed for any fish, chicken, or meat, which can be cooked right on top of the vegetables.

½ pound waxy potatoes

½ pound carrots

½ pound celery root, turnips, or rutabaga

½ pound parsnips or beets

1 medium onion, chopped

3 tablespoons olive oil, plus more as needed

Salt and freshly ground black pepper

1 tablespoon minced garlic

1 tablespoon fresh minced rosemary or sage (optional)

½ to 1 pound fish fillets, chicken breasts or tenders, boneless pork chops, or lamb ribs, divided in 4 portions (optional)

Juice of 1 lemon (optional)

1 Heat the oven to 450°F. Trim and peel the vegetables as necessary and cut them into 1-inch cubes. Mix them together in a large roasting pan or baking dish, along with the olive oil; sprinkle with salt and pepper.

2 Roast the vegetables for about 20 minutes, stirring once or twice, until they're just beginning to get tender. Add the garlic and herb if you're using it; toss to combine.

3 If you're adding fish, chicken, or meat, lay it directly on top of the vegetables, brushing with some of the oil from the pan. Sprinkle with salt and pepper and return to the oven for another 8 to 15 minutes, depending on what you're cooking, basting once or twice with the pan juices. Taste and adjust the seasoning; add the lemon juice if you like, and serve.

Braised Vegetables with Prosciutto, Bacon, or Ham

Makes: **4 servings** Time: **45 minutes**

Like meat braises and stews, the goal here is to produce the tenderness and complex flavors that come from fully softened ingredients and a luxuriously thick cooking liquid. The same technique works for any vegetable or even any combination of vegetables (I've given some specific examples in the main recipe), since the timing of doneness isn't crucial. Likewise, check out the different seasoning options in the list that follows, or experiment on your own. Serve this with whole grains (page 136) or toss with some pasta.

- 3 tablespoons olive oil
- 1 large onion or several shallots, halved and sliced
- 1/4 pound chopped prosciutto, cured or smoked ham, pancetta, or bacon (optional)
- 1 or 2 whole small dried chiles (optional)
- Salt and freshly ground black pepper
- 2 or 3 sprigs of fresh thyme or 1 sprig of rosemary or oregano
- 2 pounds any vegetable (alone or in combination), like Brussels sprouts, eggplant, carrots or parsnips, turnips or rutabaga, beets, kale or collards, string beans, or celery
- 2 cups vegetable stock (page 150), white or red wine, beer, or water
- 1/2 cup fresh chopped parsley leaves, for garnish

1 Put 2 tablespoons of the oil in a large pot or Dutch oven and turn the heat to medium. When the oil is hot, add the onion and the meat and chiles if you're using them; sprinkle with salt and pepper and cook, stirring occasionally, until the onions begin to color, about 5 minutes. Add the herb sprigs, turn the heat down a bit, and keep cooking, stirring once in a while, until the color deepens, another 5 minutes or so. Remove everything with a slotted spoon.

2 Meanwhile, trim and peel the vegetables as needed and cut them into large (at least 2-inch) chunks, or leave them whole if they're any smaller than that.

3 Return the pot to medium-high heat, add the remaining oil, and when it's hot, add the vegetables. Sprinkle with salt and pepper and cook, stirring occasionally, until they start to caramelize and brown a bit. Return the onion mixture to the pot, add the liquid, and bring to a boil. Lower the heat so that the mixture gently bubbles, cover and cook, stirring occasionally,

Braised Vegetables, Many Ways

Instead of the onion, use lots of garlic or a couple bunches of scallions, cut into big pieces.

Use peanut oil instead of the olive oil, and use tofu cubes instead of the proscuitto; try coconut milk for the liquid, and season with a splash of soy or Thai fish sauce. Garnish with chopped peanuts and cilantro.

Add a handful of chopped radishes (daikon is excellent here) during the last minute or 2 of cooking.

Instead of the meat, use chopped or sliced almonds or hazelnuts; they'll get soft and add body. For a crunchy finish, garnish with a sprinkle of whatever nut you used.

Instead of the meat, use cut-up boneless, skinless chicken thighs.

Add some citrus zest to the pot, along with the herb sprigs.

Use peanut or grape seed oil instead of the olive oil, and substitute a tablespoon or so of cumin, caraway, or mustard seeds, or a couple of caraway pods or bay leaves for the herb sprigs.

until the vegetables are as tender or as soft as you want them, anywhere from 5 to 30 minutes; just poke them with a fork every once in a while to check. When they're ready, taste and adjust the seasoning, garnish, and serve.

Grilled or Broiled Kebabs

Makes: **4 servings** Time: **45 minutes**

Kebabs are a fun way to grill different things simultaneously, while playing with marinades and spice rubs. (Start with Six Seasoning Blends You Can't Live Without, on page 143, and add some olive oil. Or simply add the spices to the salt and pepper to make a rub.) This is also the place for serving a killer salsa, which will take just a few extra minutes to throw together (see Salsa, Any Style, page 146).

Just about any vegetable that can be cut into pieces can be skewered; for meat or fish, look for beef tenderloin or sirloin; lamb or pork shoulder; chicken thighs; sturdy chunks of tuna, swordfish, halibut, or monkfish; shrimp or scallops.

About 2 pounds any vegetable or combination: eggplant, squash, peppers, onions, zucchini, cherry or plum tomatoes, mushrooms, cut into large pieces, about 1 inch thick

1 pound meat, chicken, fish, or shellfish, cut into 1-inch chunks, or whole shrimp

4 to 6 tablespoons olive or vegetable oil

Salt and freshly ground black pepper

Chopped fresh parsley or cilantro leaves for garnish (optional)

1 If you're using wooden skewers, soak them in water for at least 10 minutes. Heat a charcoal or gas grill or a broiler and put the rack about 4 inches from the heat source. If you're using charcoal or briquettes, be generous—you want a broad fire (though not the hottest fire possible).

2 Meanwhile, thread the vegetables and meat alternately on skewers, leaving a little space between pieces. Brush with the olive or grapeseed oil; sprinkle with salt and pepper. Let the kebabs sit while the fire is getting ready.

3 When the fire is hot, but still not scorching, start cooking the kebabs. Brush them with a little more oil and turn them once or twice as they cook, until they begin to brown and become tender, after 10 or 15 minutes. Put the skewers on a platter, garnish, and serve.

Pan-Cooked Grated Vegetables and Crunchy Fish

Makes: **4 servings** Time: **30 minutes**

Even the hardest vegetables—winter squash, root vegetables, and potatoes—cook relatively quickly and retain a lovely crunch if you grate them first. Pair them with lightly breaded, pan-crisped fish, and you have a colorful, fresh-tasting, and fast one-dish meal.

If you want to use zucchini or other summer squash, that's fine, too; just grate it as directed and squeeze it dry in a clean towel before cooking. And you can also make this dish with whole peas or snap peas, sliced asparagus, or ribbons of cabbage or greens.

4 tablespoons peanut or vegetable oil, plus more as needed

1 small red onion or 4 scallions, chopped

About 2 pounds winter squash, sweet potatoes, daikon, turnips, or celery root, trimmed, peeled, and grated

1 tablespoon minced ginger or garlic

1 tablespoon Fragrant or Hot Curry Powder (page 144) or to taste

Salt and freshly ground black pepper

About 1/4 cup each of cornmeal and flour, for dredging

About 12 ounces (3/4 pound) cod or catfish fillet, cut into 4 pieces

1/4 cup fresh parsley, or cilantro leaves

Lemon, lime, or orange wedges (optional)

1 Put 2 tablespoons of the oil in a large skillet (preferably cast-iron) over medium-high heat. When the oil is hot, add half the onion or scallions and the vegetables. Add the ginger or garlic and the curry powder, and sprinkle with salt and pepper. Cook, stirring, and adding a little more oil if the mixture is sticking, until the onion has caramelized and the potatoes are lightly browned, about 10 minutes; the vegetables need not be fully tender. Taste and adjust the seasoning, and transfer to a large serving platter or divide among individual plates.

2 While the vegetables are cooking, combine the cornmeal and flour on a plate along with some salt and pepper. Dredge the fish in the cornmeal mixture, pressing to make some of it stick then shaking to remove the excess.

3 Return the skillet to high heat; don't bother to wipe it out. Add the remaining oil. When it's hot, add the fish to the pan and cook, turning only once, until nicely browned on both sides and cooked through—a thin-bladed knife will meet little or no resistance when fish is done. Put the fish on top of the vegetables; garnish with the remaining onion or scallion and parsley or cilantro. Serve with citrus wedges if you like.

Other Seafood to Use for This Dish and How Long to Cook It

Shrimp (large ones will take about 2 minutes per side to cook; medium will take 3 minutes total)

Sea scallops (cook 3 to 4 minutes per side)

Salmon or halibut fillets or steaks (5 to 10 minutes total cooking time, depending on the thickness)

Tuna steaks (3 to 10 minutes total cooking time, depending on thickness and desired doneness)

Squid (cook and stir for just a couple of minutes)

Trout (cook 2 small whole fish or one large one; 5 minutes a side should do the trick)

Pan-Cooked Grated Vegetables and Sesame Fish: Instead of the cornmeal mixture, use sesame seeds to dredge the fish, pressing to make them stick to the sides. Proceed with the recipe, garnishing with cilantro and lime wedges.

Steamed Grated Vegetables with Fish: Don't bother to dredge the fish as described in Step 2, but sprinkle it with salt and pepper and have it handy when you begin the recipe; use a skillet with a tight-fitting lid. Follow the directions through Step 1, but after adding the garlic and spices, cook for just 3 minutes, then pour 1 cup of dry white wine, vegetable or fish stock (pages 150–51), or water over all and immediately top with the fish pieces. Cover and steam for 5 minutes before checking the fish for the first time. If not quite ready, cover and cook another minute or two. Serve straight from the skillet, garnished with the remaining onion or scallion and the herbs.

Orchiette with Broccoli Rabe, My Style

Makes: **About 4 servings** Time: **About 40 minutes**

I've turned the way I cook pasta upside down: Instead of making 1 pound of pasta with 1 or 2 cups of sauce, I double the sauce and cut the noodles in half. The benefit? More vegetables, maybe a little meat or seafood for flavor, and fewer refined carbohydrates. This may go against everything you've learned about cooking pasta, but it tastes wonderful, and it works with just about any recipe, except those with heavy cheese or meat sauces.

Orchiette with Broccoli Rabe is one of my favorite combinations: the slight bitterness in the broccoli rabe is balanced by the sweetness of the fennel and sausage along with a little dose of heat. Cauliflower or broccoli are also good, though they take a little longer to become tender in Step 1. For a pure vegetarian version, skip the sausage and add about 2 cups of chickpeas to the garlic along with the red pepper flakes. Or if you'd like to add a bright flavor and a bit more moisture, replace the wine with a couple of fresh chopped tomatoes.

Salt

About 1 pound broccoli rabe, trimmed and cut into pieces

1/4 cup olive oil, or more as needed

1 tablespoon chopped garlic, or more to taste

1/4 teaspoon red pepper flakes, or to taste

1 teaspoon fennel seeds (optional)

Freshly ground black pepper

1/4 to 1/2 pound sweet or spicy Italian sausage (if using link sausage, squeeze it from the casing or cut it up a bit)

1/2 cup white wine or water

½ pound dried orchiette, penne, ziti, or other cut pasta, preferably whole wheat

½ cup freshly grated Parmesan cheese (optional)

1 Bring a large pot of water to a boil and salt it. Boil the broccoli rabe until it's crisp-tender, 3 to 5 minutes, depending on the size of your pieces. Scoop the broccoli rabe out of the water with a slotted spoon or small strainer and set it aside.

2 Meanwhile, put the oil in a large skillet over medium heat. When the oil is hot, crumble the sausage into the pan, and cook, stirring occasionally to break the meat into relatively small bits and brown it, about 5 minutes. Add the garlic, red pepper flakes, and fennel seeds if you're using them, and sprinkle with salt and pepper. Continue cooking and stirring for another minute or so. Add broccoli rabe and the liquid and cook, mashing and stirring until the broccoli rabe is quite soft, 2 or 3 minutes more. Turn the heat to low to keep the sauce warm.

3 Cook the pasta in the boiling water for about 5 minutes before checking the first time. When the pasta is just tender, but not quite done, drain it, reserving about a cup of the cooking water. Toss the pasta with the sauce, along with some of the pasta water to keep the mixture from drying out. Taste and adjust the seasoning and serve immediately, with the Parmesan if you like.

Baked Ziti, My Style: Use cauliflower instead of the broccoli rabe and use ziti; grease a 2-quart baking dish with some olive oil and heat the oven to 400°F. In Step 2, before you add the cauliflower to the sausage, stir in 2 cups of chopped, peeled, and seeded tomatoes (canned are fine). Cook and stir until the water is absorbed and the mixture is saucy, then add the cauliflower and proceed. In Step 3, cook the pasta for only about 5 minutes, so that it's still quite chalky inside. When you toss the pasta with the sauce, add enough cooking water to make the mixture quite moist, but not soupy. Transfer it to the baking dish, top with the Parmesan if you like, and bake until bubbly and thickened a bit, about 20 minutes.

Super-Simple Mixed Rice, a Zillion Ways

Makes: **4 Servings** Time: **30 minutes**

Call it risotto, pilaf, Japanese rice, or soupy rice; the technique remains the same. Build a dish with rice as a base, adding ingredients from there. As you stir and cook, the rice releases starch and becomes creamy; brown rice adds a pleasant nuttiness and chew. For some ideas about varying the flavor profiles and ingredients, see the list of ideas that follows.

¼ cup dried porcini mushrooms

2 tablespoons olive oil

¾ cup short-grain brown rice

1 onion, chopped as small as you can manage

Salt and freshly ground black pepper

One 12-ounce can chopped tomatoes, with their liquid

2 cups cooked cannellini beans (page 139), or use canned

½ cup chopped fresh basil or parsley leaves, plus more for garnish

½ cup grated Parmesan cheese (optional)

1 Soak the porcini in hot water to cover. Put the olive oil in a large pot over medium heat. When the oil is hot, add the rice and cook, stirring constantly, until it's shiny and a little translucent, just a minute or so. Add the onions, sprinkle with salt and pepper, and continue to cook, stirring occasionally, until softened, about a minute. Add enough water to cover by about half an inch.

2 Bring to a boil, lower the heat to a bubble, and cook, stirring occasionally, until the rice starts to become tender, about 10 minutes. By now the porcini should be soft; chop them roughly and pour their water into the rice, being careful to leave some water behind to trap the sediment. Add the tomatoes and mushrooms to the rice and continue to cook and stir occasionally, until the tomatoes break down, about another 10 minutes; add more water if needed to keep the mixture a little soupy.

3 When the rice is tender but retains some bite on the inside, add the beans. Continue to cook, stirring occasionally, until the mixture is no longer soupy but not yet dry. Stir in the basil, and the cheese if you're using it; taste and adjust seasoning and serve, garnished with a little more fresh herb.

Some Easy Variations for Mixed Rice:

Rice and Peas (*Risi e bisi*): Omit the mushrooms (or not). If you do omit them, add another cup of water in Step 2. Likewise with the tomatoes: keep them or not. Instead of adding the beans in Step 3, add 2 cups of peas (frozen are fine).

Chile Mixed Rice: Toss a seeded chipotle or pasilla chile into the rice mixture with the onion, along with a clove or two of chopped garlic. Use black beans instead of the cannellini and cilantro instead of the basil. Omit the cheese and stir in a cup or so of cooked chicken or pork if you like.

Japanese Mixed Rice: Substitute dried shiitakes for the porcini. Instead of the olive oil, use a mixture of half peanut oil and half sesame oil. Omit the tomatoes and beans and stir in other vegetables like bean sprouts, asparagus, or chopped bok choy. At the same time add a few shrimp or cubes of tofu if you like. Keep the mixture moist with water, and at the very end add a big splash of soy sauce instead of the cheese. Garnish with chopped scallions or chives instead of the basil.

Coconut Mixed Rice: Omit the porcini and use peanut or grape seed oil instead of olive oil. When you're cooking the onion with

Substituting Brown Rice for White Rice in Any Recipe

This is the simplest trick ever, and it works even for risotto. Bring a pot of water to a boil (as you would for pasta) and salt it. Add the amount of brown rice specified in whatever recipe you have that calls for white rice. (If the recipe calls for long-grain rice, use long-grain brown rice, and so on.) Boil the rice, stirring occasionally, for about 12 minutes. Drain and use within an hour or so, or put in the fridge for up to a day. Whenever you're ready, begin the recipe as usual, using the parboiled brown rice instead of white rice.

the rice, add some chopped eggplant, one tablespoon of minced ginger, and another of Hot or Fragrant Curry Powder (page 144). When you add the tomatoes, add a cup of coconut milk. Use chickpeas instead of the cannellini and garnish with cilantro instead of the basil. Omit the cheese and sprinkle with chopped pistachios if you like.

Paella

Makes: **4 to 6 servings** Time: **About 45 minutes**

Paella often looks like a big deal, but in its soul it is rice with good stuff in it, and that stuff can be anything. (The common restaurant paella is just one interpretation.) This version relies on good tomatoes, but you can use eggplant, root vegetables, or winter squash, or improvise in other ways: substitute a few scallops or mussels for the sausage, for example. For a vegetarian version, skip the sausage and shrimp; toss slices of eggplant, zucchini, mushroom caps, or carrots in some olive oil; and set them on top of the rice to roast.

To use white rice here, reduce the oven time by about half.

- 3½ cups any stock (page 150), or water, plus more if needed
- Pinch saffron threads (optional)
- 1 pound ripe tomatoes, cored and cut into thick wedges
- Salt and freshly ground black pepper
- 3 tablespoons olive oil
- 1 medium onion, chopped
- 1 tablespoon tomato paste
- 2 teaspoons Spanish *pimentón* (smoked paprika), or other paprika
- 3 ounces Spanish chorizo or other cooked or smoked sausage, cut into ¼-inch pieces (optional)
- 2 cups short-grain brown rice
- 6 large peeled shrimp; or 2 boneless chicken thighs, cut into ½-inch chunks (optional)
- 1 cup fresh or frozen peas
- Chopped fresh parsley leaves for garnish

1 Heat the oven to 450°F. Warm the stock, and the saffron if you're using it, in a small saucepan. Put tomatoes in a medium bowl, sprinkle with salt and pepper, and drizzle them with 1 tablespoon olive oil. Toss to coat.

2 Put the remaining oil in a 12-inch ovenproof skillet over medium-high heat. When the oil is hot, add the onion, sprinkle with salt and pepper, and cook, stirring occasionally, until it softens, 3 to 5 minutes. Stir in tomato paste, paprika, and chorizo if you're using it, and cook for a minute more.

3 Add the rice and cook, stirring occasionally, until it's shiny, another minute or two. Carefully add the warm stock and stir until just combined, then stir in the shrimp and peas.

4 Put the tomato wedges on top of the rice and drizzle with the juices that accumulated in the bottom of the bowl. Put the pan in the oven and bake, undisturbed, for 30 minutes. Check to see if the rice is nearly dry and just tender. If not, return the pan to the oven for 5 minutes, check again, and repeat if necessary. If the rice looks too dry at any point, but still isn't quite done, add a small amount of stock or water. When the rice is ready, turn off the oven and let it sit for at least 5 and up to 15 minutes.

5 Remove the pan from the oven and sprinkle with parsley. If you like, put the pan over high heat for a few minutes to develop a bit of a bottom crust before serving.

Bulgur Pilaf with Vermicelli, and Meat or Cauliflower

Makes: **4 servings** Time: **30 minutes**

The traditional version of this recipe makes a glorious side dish, but if you add protein and vegetables it becomes a meal in itself. Serve with a big salad or just before the resting time in Step 3, toss in any chopped tender green, like spinach or arugula, and let the leaves steam a bit right in the dish. If you don't have vermicelli on hand, this is a great opportunity to use up any pasta odds and ends like penne, ziti, or shells; put whatever you have in a plastic bag and break them up with a skillet or rolling pin.

2 tablespoons olive oil

¼ to ½ pound ground lamb, beef, turkey, or chicken; or 1 small head cauliflower, cored and roughly chopped

Salt and freshly ground black pepper

1 pound of any fresh mushrooms, sliced

2 medium onions or 1 large onion, chopped

½ cup vermicelli, preferably whole wheat, broken into 2-inch-long or shorter lengths, or other pasta

1 cup coarse- or medium-grind bulgur

1 tablespoon tomato paste (optional)

2¼ cups vegetable stock (page 150), or water, heated to the boiling point

¼ cup chopped fresh parsley leaves, for garnish

1 Put the oil in a large skillet or saucepan that can later be covered and turn the heat to medium. Add the meat or cauliflower, sprinkle with salt and pepper, and cook, stirring occasionally to break it up, until browned all over, about 10 minutes.

Remove from the pan and spoon off all but a couple tablespoons of the fat.

2 Put the pan over medium-high heat. Add the mushrooms and onions; cook, stirring, until everything is soft, about 5 minutes. Add the vermicelli and the bulgur and cook, stirring, until coated with butter or oil. Return the meat or cauliflower to the pan and add all the remaining ingredients. Turn the heat to low, and cover. Cook for 10 minutes.

3 Turn off the heat and let the mixture sit for 15 minutes more. Taste and adjust the seasoning, fluff with a fork, and serve, garnished with a sprinkling of parsley.

Bean and Vegetable Chili

Makes: **6 to 8 servings** Time: **About 45 minutes, with cooked or canned beans**

A traditional chili made special. Kidney beans or pinto beans are a good starting point, but also consider chickpeas, black beans, cannellini, navy beans, cranberry beans, marrow beans, borlotti, or even lentils (which break down and lend a terrific meatiness). Any or all of these will give you a creamy, earthy bowl of chili.

The same flexibility goes for the vegetables: You can swap the eggplant, zucchini, and mushrooms for almost anything you have in the fridge. And don't hesitate to toss in a cup of cooked grains like wheat berries, wild rice, or barley.

3 tablespoons olive oil

½ pound ground beef, pork, turkey, or chicken (optional)

Salt and freshly ground black pepper

1 whole onion, unpeeled; plus 1 small onion, minced

1 tablespoon minced garlic

1 or 2 small eggplants (peeled if you like), cubed

1 medium zucchini, chopped; or use more eggplant

1 or 2 carrots, chopped

1 cup quartered mushrooms or a handful of rinsed dried porcini

1 fresh or dried hot chile, seeded and minced, or to taste (optional)

1 teaspoon ground cumin, or to taste

1 teaspoon minced fresh oregano leaves or ½ teaspoon dried oregano

1 cup peeled, seeded, chopped tomato (canned is fine; don't bother to drain)

- 4 cups cooked kidney beans or pinto beans (page 139), liquid reserved (canned are fine)
- 2 cups vegetable stock (page 150), or water, more or less as needed (optional)
- Chopped fresh cilantro or parsley for garnish

1 Put the oil in a large pot over medium heat, and when the oil is hot add the meat if you're using it. Sprinkle with salt and pepper and cook, stirring frequently, until it is well browned all over, about 10 minutes. Remove the meat from the pan and drain off all but 3 tablespoons of the fat. (If you're skipping the meat, put the oil in the pan and start the recipe here.)

2 Return the pan to the stove, this time over medium-high heat. Add the onion and garlic and cook and stir until just softened, about a minute. Add the vegetables, sprinkle with salt and pepper, and cook, stirring occasionally, until they begin to soften and become fragrant, adjusting the heat so that nothing scorches. After about 10 minutes, the vegetables should start to caramelize a bit and dry out.

3 Add the seasonings and stir, then add the tomatoes and beans, with enough of their liquid to submerge everything (use some stock or water if you don't have enough). Bring the mixture to a boil and cook, stirring occasionally and adding more liquid if necessary, until the beans are very tender and the flavors have mellowed, about 15 minutes. Taste, adjust the seasoning, and garnish with cilantro. Serve with brown rice, crackers, tortilla chips, or whole grain bread.

Bean and Vegetable Chili with Tofu: Instead of the ground meat, crumble as much as 1 pound of tofu into the hot oil in Step 1.

Cassoulet with Lots of Vegetables

Makes: **4 to 6 servings** Time: **40 minutes**

Cassoulet is one of the best of the myriad of traditional European dishes that combine beans and meat to produce wonderful rich, robust stews. This recipe maintains that spirit, but is much faster, easier, less expensive, and more contemporary, emphasizing the beans and vegetables over meat. (That probably makes it *more,* not less, traditional, since meat was always hard to come by before the mid-twentieth century.)

The main recipe starts with already cooked beans or canned beans and is ready relatively fast. To begin with dried beans, see the variation; it takes more time, but the results are even better.

2 tablespoons olive oil

1 pound Italian sausages, bone-in pork chops, confit duck legs, or duck breasts, or a combination

1 tablespoon chopped garlic

2 leeks or onions, trimmed, washed, and sliced

2 carrots, peeled and cut into 1-inch lengths

3 celery stalks, cut into ½-inch pieces

2 medium zucchinis or 1 small head green cabbage, cut into ½-inch pieces

Salt and freshly ground black pepper

4 cups chopped tomatoes, with their juice (canned are fine)

¼ cup fresh chopped parsley leaves

1 tablespoon fresh chopped thyme leaves

2 bay leaves

4 cups cooked white beans (canned are OK), drained and liquid reserved in any case

2 cups stock, dry red wine, bean cooking liquid, or water, plus more as needed

1/8 teaspoon cayenne pepper, or to taste

1 Heat the olive oil in a large saucepan over medium-high heat, add the meat, and cook, turning as needed, until the meat is deeply browned on all sides, about 10 minutes. Remove from the pan and drain off all but 2 tablespoons of the fat.

2 Turn the heat to medium and add the garlic, leeks or onions, carrots, celery, and zucchini or cabbage; and sprinkle with salt and pepper and cook until softened, about 5 minutes. Add the tomatoes, their liquid, the reserved meat, and the herbs and bring to a boil. Add the beans; bring to a boil again, stirring occasionally, then reduce the heat so the mixture bubbles gently but continuously. Cook for about 20 minutes, adding the liquid when the mixture gets thick and the vegetables are melting away.

3 Fish out the meat and remove the bones and skin as needed. Chop into chunks and return to the pot along with the cayenne. Cook another minute or two to warm through, then taste and adjust seasoning if necessary and serve.

Slow-Cooked Cassoulet: Start with dried beans. After browning the meat in Step 1, leave it in the pan and add ½ pound dry white beans (they'll cook faster if you soak them first; see page 139) and enough water or stock to just cover. Bring to a boil, then reduce the heat and cook, stirring occasionally, for about an hour. Meanwhile, in a separate pan with another 2 tablespoons of olive oil, cook the vegetables as directed in Step 2. Add them to the pot of beans along with the tomatoes and herbs. Bring to a boil, then reduce the heat to a gentle bubble and cook, stirring occasionally, until the beans are tender, adding more liquid as necessary to keep them moist. This will take anywhere from another 30 to 60 minutes, depending on the age of your dried beans.

Chickpea Stew with Roasted Chicken

Makes: **4 servings** Time: **About 1 hour with cooked chickpeas**

Chickpeas are quite possibly the most flavorful legume; for that reason, the chicken in this recipe is truly optional—with or without it, you'll get a hearty, wonderful dish. Substitute lamb chops for the chicken if you like, or simplify the recipe by replacing the chicken with a few ounces of cooked sausage or chorizo (or even shrimp or squid) in Step 4 when you combine the vegetables and the chickpeas.

- 4 cups cooked chickpeas (page 139) or canned chickpeas, favas, or lima beans, drained but reserving the liquid
- 2 cups bean cooking liquid, any stock (pages 150–51), or water
- Salt and freshly ground black pepper
- 4 tablespoons olive oil
- 4 chicken pieces, about 1 pound; preferably legs or thighs (optional)
- 1 or 2 small eggplants or medium zucchinis, chopped
- 1 pound fresh mushrooms (any kind), trimmed and sliced
- 1 large onion, chopped
- 1 celery stalk, chopped
- 1 carrot, chopped
- 1 tablespoon minced garlic
- 1 teaspoon minced fresh ginger
- 1 teaspoon ground coriander
- 2 teaspoons ground cumin
- 2 cups peeled, seeded, chopped tomato (canned is fine; don't bother to drain)
- Chopped fresh cilantro or parsley leaves for garnish

1 Heat the oven to 400°F. Warm the beans in a large pot with the liquid; sprinkle with some salt and pepper. Adjust the heat so the mixture bubbles slowly.

2 Put 3 tablespoons olive oil in a large, deep skillet over medium heat, add the chicken, sprinkle with salt and pepper and brown well on all sides, for about 10 minutes. Transfer the chicken to a small roasting pan and set aside.

3 Pour off all but 3 tablespoons of the fat and return the pan to medium heat. Add the eggplant or zucchini and the mushrooms, sprinkle with salt and pepper, and cook, stirring occasionally, until tender, 15 to 20 minutes. Remove from the pan with a slotted spoon and set aside. Put the chicken in the oven.

4 Add another tablespoon of oil to the skillet. Still over medium heat, add the onion, celery, and carrot. Cook, stirring occasionally, until the vegetables are softened, about 10 minutes. Add the garlic, ginger, coriander, cumin, and tomato and cook for 5 minutes more, stirring occasionally and scraping the bottom of the pan to loosen any brown bits. Add the mixture to the simmering beans along with the reserved eggplant or zucchini and mushrooms.

5 When the chicken has cooked for about 15 minutes, check for doneness (the juices will run clear if you make a small cut in the meat near the bone). When it is ready, remove it from the oven. When the vegetables are tender, put the chickpeas and the vegetables on a large, deep platter; top with the chicken, and drizzle with its juices; then garnish, and serve.

Chicken Not Pie

Makes: **4 Servings** Time: **30 minutes**

Can chicken potpie still be considered comfort food if you leave out the crust? I think so, especially since this version includes lots of creamy potatoes, all the familiar vegetables, and a couple of surprises to boot. Frozen peas are a good substitute when fresh peas aren't available; but if asparagus is out of season, try ribbons of cabbage or chard. Serve this with Easy Whole Grain Flatbread (page 224), Almost No-Work Whole Grain Bread (page 156), or Hybrid Quick Bread (page 154). It's also good over plain brown rice or whole wheat fettuccini.

¼ cup olive oil

2 leeks, washed well and diced, including some of the green part

Salt and freshly ground black pepper

1 cup dry white wine or water

1 cup vegetable stock (page 150) or more water

½ teaspoon chopped fresh thyme or tarragon leaves or a good pinch dried thyme or tarragon

2 boneless chicken breasts (1 whole breast, split in half), or 4 whole chicken tenderloins

2 or 3 large all-purpose potatoes (like redskin or Yukon Gold), peeled if you like and cut into 1-inch cubes

2 medium carrots or parsnips, cut into coins

½ pound sugar snap peas or snow peas, trimmed and strings removed if necessary; or 1 cup shell peas (frozen are fine)

½ pound asparagus, trimmed and cut into 1-inch pieces

2 tablespoons freshly squeezed lemon juice

Chopped fresh parsley leaves for garnish

1 Put half of oil in a large skillet over medium heat. When the oil is hot, add the leeks, sprinkle with salt and pepper, and cook, stirring occasionally, until softened, about 5 minutes. Add the wine, stock, and herb; bring to a boil, and let bubble for a minute or two.

2 Add the chicken, turn the heat down to medium-low, cover, and simmer until the meat is barely cooked through, 5 or 6 minutes. Remove the chicken.

3 Add the potatoes and bring to a boil; reduce the heat so the liquid bubbles enthusiastically; and cook until the potatoes are almost tender, about 5 minutes. Stir in the carrots or parsnips and cook for another couple of minutes. By now the liquid should be thickening; if not, turn the heat up and cook another couple of minutes, stirring to prevent the vegetables from sticking. Add the remaining oil gradually, stirring vigorously with the back of a spoon as you do so.

4 Add the peas and asparagus to the pot. Cook, stirring occasionally, until the vegetables are brightly colored and just tender, about 3 minutes. Chop or slice the chicken and return it to the pot, along with any juices that have accumulated and the lemon juice. When warmed through, taste and adjust the seasoning. Serve in shallow bowls, garnished with the parsley.

Stuffed Chicken Breasts with Pan-Grilled Corn

Makes: **4 to 8 servings** Time: **45 minutes**

Everyone likes something fancy on occasion, and though most recipes for stuffed chicken are a pain, this one uses whole boneless breasts to enclose the filling, requiring only three ties around each to hold things together. (And you use just one skillet!)

Think of an inside-out fajita, stuffed with peppers and onions, then cooked and sliced. And you can try almost anything for the stuffing: leftover cooked vegetable, grains, or beans; nuts and plumped dried fruit; reconstituted dried mushrooms or dried tomatoes; or even slices of fresh fruit, like plums or peaches; keep the filling to less than a cup. (You can also use a different vegetable instead of the corn, and treat it the same way.)

- 3 tablespoons olive oil
- 1 medium onion, chopped
- Salt and freshly ground black pepper
- 1 medium red bell pepper, cored, seeded, and chopped
- 4 boneless chicken breasts (about 1½ pounds)
- 6 ears fresh corn, shucked; or use about 4 cups of frozen corn
- 1 fresh chile, like serrano or jalapeño, seeded if you like, and chopped
- 1 tablespoon minced garlic
- 2 tablespoons sherry or white wine vinegar, or the juice of 1 lime
- ½ cup chopped fresh cilantro leaves

1 Heat the oven to 350°F. Cut six 12-inch pieces of butcher twine. Put 2 tablespoons of the olive oil in a large ovenproof skillet over medium heat. When the oil is hot, add the onions, sprinkle with salt and pepper, and cook, stirring occasionally, until softened, fairly dry and beginning to color, about 5 minutes. Add the peppers and cook, stirring frequently until the mixture softens and comes together a bit, another 3 minutes or so.

2 Spread 2 chicken breasts out on a work surface so that the side where the bones were is faceup; flatten them a bit with the bottom of a pot or the palm of your hand. Spread about ½ cup of the onion mixture on top of each breast, then top with the remaining breasts, end to end, so that the tapered side is on the same side as the rounded side (this ensures a tidy roll and even cooking). Tie each stuffed breast in three places with twine. Sprinkle all sides with salt and pepper and roll them around in the same pan you used to cook the vegetables, coating them in whatever oil and juices remain in the pan. Transfer to the oven and bake, turning once, until the chicken is cooked through and opaque, 25 to 30 minutes.

3 Remove the chicken from the pan and tent with foil. Set the pan over high heat and add the last tablespoon of oil. When the oil is hot, add the corn, chile, and garlic; let sit for a moment. As the corn browns, use a spatula to toss the corn so that each kernel is deeply browned on at least one side. Remove from the heat, then sprinkle with salt and pepper and stir in the vinegar or lime juice, a tablespoon or two of water, and the cilantro, scraping to stir up any browned bits from the pan. Remove the string from the chicken and slice crosswise into thin or thick pieces. Serve, with a big spoonful of the corn on the side.

Eggplant and Chicken Parmesan

Makes: **4 to 6 servings** Time: **About 1 hour**

Eggplant Parmesan is a lot of work. Here's a more straightforward version, with the vegetables and meat grilled or broiled instead of breaded and fried. You can skip the chicken if you like, and add other vegetables, like zucchini and portobello mushrooms; just grill them and layer on top of the eggplant and before the cheese.

For a simple vegetable gratin, omit the tomato sauce and layer any cooked vegetable you like (asparagus, broccoli, cauliflower, artichoke hearts, potatoes, fennel, leeks, spinach, onions, celery root, parsnips, Jerusalem artichokes, winter squash, or sweet potatoes) with the cheese (Gruyère and Swiss are nice alternatives). Finish with the seasoned bread crumb topping for a most excellent crust.

2 or 3 eggplants (about 2 pounds total), unpeeled, and cut crosswise into ½-inch slices

Salt

1 tablespoon olive oil, plus more for brushing

Freshly ground black pepper

About ½ pound boneless, skinless white meat chicken (breast, cutlets, or tenders), pounded to uniform thickness if necessary and blotted dry

4 cups All-Purpose Tomato Sauce (page 147)

1 cup grated Parmesan cheese, plus more if you like

About 30 fresh basil leaves

2 ounces grated or torn mozzarella cheese (optional)

1 cup bread crumbs

1 If the eggplant is particularly large or full of seeds, sprinkle it with salt and set in a colander for at least 15 and up to 60

minutes. Rinse and pat dry. Heat the oven to 400°F. Heat a charcoal or gas grill, or the broiler, and move the rack to about 4 inches from the heat source. (You can also use a stovetop grill pan here, heated over medium-high heat.)

2 Brush the eggplant lightly on both sides with some oil and sprinkle with salt (if you didn't salt it earlier) and pepper. Grill or broil until browned on both sides, turning once or twice and brushing with more oil if the eggplant looks dry. The idea is to keep the eggplant cooking steadily without burning, so adjust the heat and position as needed. The eggplant is usually ready in somewhere between 5 and 10 minutes. When done, set eggplant slices aside.

3 Cut the chicken so you have 8 or so large pieces. Pound or press them a bit so they're evenly flat. Brush them all over with some oil and sprinkle with salt and pepper. Grill or broil the chicken, turning once, no more than 3 minutes per side (to check for doneness, cut into a piece with a thin-bladed knife; the center should still be slightly pink). Set the chicken aside.

4 Lightly oil a 2-quart baking dish, then spoon a layer of the tomato sauce into the bottom. Top with a layer of eggplant, then a sprinkling of Parmesan, then a layer of chicken, and finally a few basil leaves. Repeat until all the ingredients are used. (There will probably be sauce left over; warm it up to pass at the table.) Toss the remaining Parmesan with the bread crumbs, and the mozzarella if you're using it. Drizzle with 1 tablespoon of olive oil, sprinkle with salt and pepper, then toss again. Spread the bread crumb mixture evenly on top of the mozzarella. Bake for 30 to 35 minutes, or until the dish is bubbling hot. Serve hot or warm.

Roasted Herb-Stuffed Vegetables

Makes: **4 servings** Time: **50 minutes with cooked grains**

There's something not only satisfying but lovable about stuffed vegetables, which look appealing no matter what you use for filling. They're also a good opportunity to experiment with different grains and beans: brown rice, quinoa, couscous, lentils, white beans, wild rice, and wheat berries are all good choices. I often use leftover grains, and of course you can throw a little meat in there if you crave it—ground pork is most traditional, but just about anything will work; see the list below. In any case, add a salad and you've got the perfect weeknight dinner.

If you can create a hollow space in the interior of a vegetable, you can stuff it; the best and easiest options are eggplants, bell peppers, tomatoes, zucchini, and summer squash. Acorn and other winter squash work as well, but you'll need to cook these first. In general, figure about 8 ounces of vegetable per person; so for four people, you'd use 4 small eggplants or 2 medium, 4 fat straight zucchini or yellow squash, 4 large tomatoes or bell peppers, or 2 acorn squash.

About 2 pounds of stuffable vegetables (see above)

¼ cup olive oil, plus more as needed

Up to 2 cups of any leftover bean or vegetable (optional)

2 cups cooked grains (page 136)

1 tablespoon minced garlic, or more

1 cup chopped fresh parsley or basil leaves, plus more for garnish

1 teaspoon fresh thyme or rosemary leaves, or ½ teaspoon tarragon

Salt and freshly ground black pepper

1 Heat the oven to 375°F. For eggplant and squash, cut in half and use a spoon to scoop out a cavity. For tomatoes and peppers, slice off a lid to create a container and scoop out the insides. Discard the seeds and stringy pulp, but save any edible flesh and roughly chop it.

2 If you have any edible vegetable left from hollowing out the cavity, or are using leftover beans or anything else, put 2 tablespoons of the olive oil in a large skillet over medium-high heat. When the oil is hot, add the vegetable bits, sprinkle with salt and pepper, and cook, stirring occasionally, until the pieces are beginning to get tender and are relatively dry. If you're using only the grains and herbs for stuffing, skip to Step 3.

3 Mix together the cooked grains, garlic, herbs, and any other ingredients you like (see list below for suggestions). Sprinkle the grain mixture and the inside of the vegetables with salt and pepper, stuff them and, for tomatoes and peppers, replace the top slices. Spread half the remaining olive oil in a shallow roasting pan that will allow for a little room between the vegetables and put them in the pan. Sprinkle the tops with salt and pepper and put the roasting pan in the oven.

Great Additions to Grain Stuffings

Crumbled cooked sausage, diced chorizo, or any minced leftover (or fresh) meat or fish

Chopped nuts (pecans, pine nuts, walnuts, almonds)

Olives or capers

Raisins, currents, or dried cranberries

Grated Parmesan cheese or crumbled feta

4 Roast the vegetables for 20 to 40 minutes, until the flesh is tender and the stuffing is hot; the cooking time will vary depending on the vegetable. Serve hot, warm, or at room temperature, drizzled with the rest of the olive oil and garnished with the remaining herbs.

Stuffed Acorn and Other Winter Squash: Halve the squash, scrape out the seeds, and rub the inside with some olive oil; roast, cut side down, in a 375°F oven for 25 minutes before stuffing as described in Step 3. Proceed with the recipe.

Modern Bouillabaisse

Makes: **4 servings** Time: **About an hour**

Traditionally, people living on the Mediterranean coast made bouillabaisse using whatever scrap fish and market produce they had handy. Over the years, as people have become richer and fish more widely available, the vegetables have become almost an afterthought. In a way this recipe is a return to tradition, offering plenty of flexibility with the fish—firmer fish is usually better, so it holds together, but any fish will "work"—and plenty of vegetables. You can always add a few clams or mussels to the pot, or any other fish you like. Note that the cooking time here is largely for preparation; the stew simmers for only a few minutes.

Roasted Red Pepper Sauce is akin to the authentic *rouille*, whether or not you choose to add any mayonnaise.

1 recipe Roasted Red Pepper Sauce (page 153) (optional)

2 tablespoons mayonnaise, preferably homemade (optional)

1 tablespoon minced garlic, plus a little more for the *rouille* if you're making it

2 tablespoons olive oil

1 fennel bulb, thinly sliced

2 leeks, white and tender green parts, trimmed and cut into coins (or use onions)

Zest from 1 orange

Big pinch saffron (optional)

1 dried hot chile, or a pinch of cayenne, or to taste

1 sprig fresh tarragon (optional)

2 cups chopped tomatoes (canned are fine; drain them first)

About 1½ pounds small red or white potatoes, peeled if you like and cut into wedges

About 1 pound almost any seafood, like monkfish, cod, scallops, squid, or shrimp; peeled, skinned, boned, and cut into chunks as needed

2 carrots or parsnips, cut into coins

2 stalks celery (with the leaves if you like), cut into chunks

½ pound sugar snap peas or snow peas (optional)

2 cups vegetable, shrimp, or fish stock (pages 150–51), dry white wine or water, plus more as needed

½ cup roughly chopped parsley leaves, or use chopped chervil or fennel fronds if you like

Salt and freshly ground black pepper

1 or 2 whole grain baguettes, cut crosswise into slices and toasted if you like (optional)

1 If you're making the sauce (*rouille*), combine the red pepper sauce with a pinch of minced garlic and the mayonnaise if you're using it. Set aside for the flavors to blend.

2 Put the olive oil in a large pot or Dutch oven over medium heat. Add the fennel bulb, leeks, 1 tablespoon of garlic, and zest and cook, stirring occasionally, until softened, about 3 minutes. Add the saffron if you're using it, the chile or cayenne, and the tarragon if using and cook for about a minute. Add the tomatoes and potatoes and cover.

3 After about 5 minutes, lift the lid and stick a fork in the potatoes; if they're not yet beginning to get tender, cover and cook another couple of minutes. Try sticking the potatoes again; when the fork meets with just a little resistance, add the fish, carrots or parsnips, celery, snap peas or snow peas if you're using them, and stock, adding enough extra to just cover the fish and vegetables.

4 Bring to a boil, then cover and turn off the heat. Let the pot rest for about 5 minutes; the vegetables you just added should be crisp-tender, and the fish should be opaque and cooked through (if not, return the pot to a simmer again for a couple of minutes). Stir in the parsley, taste and adjust seasoning, and serve with the bread and sauce (if you made it) passed at the table.

Meat-and-Grain Loaves, Burgers, and Balls

Makes: **4 to 8 servings or more, for appetizers**

Time: **About an hour, or less if you start with preground meat, cooked grains, or soaked bulgur**

People throughout the world combine meat and grain to get a little extra mileage out of a precious foodstuff, and there's an obvious economic benefit. Start with whatever raw meat you like, remembering that a little bit of fat helps the flavor and texture.

The recipe here serves four to six as a main course, eight or more for appetizers. You can shape and freeze the uncooked mixture for up to a month or so, then cook it directly from its frozen state in the microwave or oven. Or cook up a batch and freeze the leftovers for up to several weeks.

2 tablespoons vegetable oil, plus more as needed

Salt

1 pound fresh spinach leaves

1 pound boneless chicken or turkey thighs, beef chuck or sirloin, or pork or lamb shoulder, excess fat removed; or use already ground meat

1 small onion, chopped as small as you can manage

2 cloves garlic, chopped as small as you can manage

Pinch cayenne

1 teaspoon ground cumin or 1 tablespoon chili powder (page 143)

Freshly ground black pepper

1 egg

2 cups soaked and drained bulgur (page 137), or other cooked grains like cracked wheat, steel-cut oats, or whole wheat couscous

1 Heat the oven to 400°F. Grease a loaf pan, rimmed baking sheet, or large roasting pan with 2 tablespoons of the oil. Bring a large pot of water to boil and salt it, and fill a bowl with ice water. Put the spinach into the boiling water for about 30 seconds. Drain and immediately plunge into the ice water. Drain, squeeze tightly to dry thoroughly, and roughly chop. Put the spinach in a large bowl. If you're using preground meat, add it to the bowl and skip to Step 3.

2 If you're using whole pieces of meat, cut them into large chunks and put in a food processor. Pulse several times to process until ground but not pureed, stopping the machine and scraping down the sides if necessary. Transfer to the large bowl.

3 Add the onion, garlic, and spices and sprinkle with salt and pepper. Stir, add the egg and bulgur or grains, then stir until thoroughly combined (a rubber spatula or your hands are ideal here). Transfer the mixture to the loaf pan or shape into a free-form loaf, burgers, or balls and put on the baking sheet or in the roasting pan. Transfer to the oven and roast until firm and browned all over: A loaf will take about 50 minutes; burgers and balls with take 20 to 30, depending on their size (carefully turn them once or twice for even cooking).

Fish or Shrimp Loaves, Burgers, or Balls: Instead of the boneless meat, use 1 pound of raw tuna, salmon, cod, halibut, catfish, or shrimp; clean, bone, and shell it as needed. If you like, instead of the cumin or chili powder, try 1 tablespoon Fragrant or Hot Curry Powder (page 144). Proceed with the recipe.

Bean-and-Grain Loaves, Burgers, or Balls: Lentils, pinto beans, and black beans work best, but any bean will do. Instead of the boneless meat, use 2 cups cooked or canned beans, mashed with enough of their liquid to keep them moist. Proceed with the recipe.

Savory Vegetable and Grain Torta

Makes: **4 to 8 servings** Time: **2 hours plus resting time**

Layers of vegetables on a whole grain crust make this torta not only delicious but impressive enough to serve to guests. Choose whatever grain you like: quinoa, cracked wheat, farro, steel-cut oats, kasha, and millet are all good. The same holds true for the vegetables; I like the Mediterranean flavors in this version, but you can also try thinly sliced and par-cooked butternut squash, potato, sweet potato, caramelized leeks, oven-roasted tomatoes or peppers, even chiles. For a touch of meat, sprinkle on a thin layer of cooked and crumbled Italian sausage or chopped pancetta.

2 medium yellow onions, halved and thinly sliced

About 3⁄4 cup olive oil

2 medium eggplant, unpeeled, cut crosswise into 1⁄2-inch slices

4 medium zucchinis, cut into slices 1⁄4 inch thick

Salt

Freshly ground black pepper

4 cups cooked grain (page 136)

20 or so fresh basil leaves, some chopped or torn for garnish

Freshly grated Parmesan cheese for garnish (optional)

1 Heat the oven to 400°F. Put the onions in a large dry skillet with a lid over medium heat. Cover and cook, stirring infrequently, until the onions are dry and almost sticking to the pan, about 20 minutes. Add 2 tablespoons of the oil and cook, stirring occasionally, until the onions brown, another 10 to 15 minutes. Set aside.

2 Meanwhile, smear 2 (or more) baking sheets with 2 tablespoons oil each. Lay the eggplant slices on one sheet and the zucchini on the other in a single layer. (You may need to work in batches, cooking the eggplant on the 2 baking sheets, then cooking the zucchini.) Sprinkle with some salt and pepper and drizzle or brush each with another couple tablespoons of oil. Roast until the eggplant and zucchini are soft, about 15 minutes for the zucchini, 20 or so for the eggplant.

3 Coat the bottom and inside the ring of an 8- or 9-inch springform pan with some oil. Press half of the cooked grains into the bottom of the pan to form an even crust, about ½ inch thick, covering the bottom completely. Layer a third of the eggplant slices on top of the grain crust (trim the eggplant pieces if necessary to fit), then layer half the zucchini, half of the caramelized onions, and some of the basil, sprinkling each layer with a bit of salt and pepper; repeat the layers, pressing down gently on each one, and ending with eggplant. Spread the remaining grain on top and press with a spatula or spoon to make the torta as compact as possible. Sprinkle the top with Parmesan cheese if you're using it.

4 Put the torta in the oven and cook until it is hot throughout and the cheese is crusty, about 30 minutes. Let sit for about 5 minutes before carefully removing the outer ring of the pan. Then let cool for another 10 minutes before cutting into wedges. Garnish with the remaining basil and serve.

Other Dishes in the Book You Can Eat for Dinner

Desserts

There is no reason for an eating style that excludes dessert. While chocolate cake with ice cream is not going to cut it on a daily basis, the model here is mostly sane food with the occasional binge. This chapter represents exactly that.

A couple of things to bear in mind. One, the number of treats you allow yourself daily can fluctuate according to how you feel; remember, it's the long run that matters, not any given day. If you've already indulged in a bag of chips in the afternoon and a steak in the evening, you might reconsider the ice cream sundae; on the other hand, if your day has been rigorously lean, go for it.

Two, consider the composition and size of a dessert. We all agree that desserts usually should be sweet, and these fit that description. What's not essential to desserts is very high fat, super-high calories, junk, or massive portions. (There's a big difference between a huge slice of cake and a sliver. Just reminding you.)

Sometimes dessert is a good pear, or slices of fruit with a sweet topping. Sometimes, yes, it's that cake. Mix them up, keep them under control, and not only will you be satisfied, you'll be fine.

Here are a few of my favorites. There are many options, from the marvelous No-Bake Fruit Tarts to Chocolate Fondue to a more or less classic Rice Pudding. None will make you feel that you're making a sacrifice.

No-Bake Fruit Tarts

Makes: **Enough for 8 individual 3-inch tarts** Time: **20 minutes**

Unlike flour-based crusts, this one—a combination of nuts and dried fruit making for a sweet, chewy consistency that's a perfect base for fresh fruit—can't be overworked, and you never need to turn on the oven. I sometimes add a thin layer of melted dark chocolate in between the crust and the fruit, both to help anchor the fruit and as a little decadent surprise; but you can skip it.

For the topping, berries are a natural choice, but try sliced apricots, plums, peaches, figs, or even kiwi. Dried cherries in the base with fresh cherries on top is lovely; the same goes for apricots or figs.

- About 4 cups fresh fruit, trimmed, cored, peeled, and sliced as needed
- 1 tablespoon sugar
- 1/2 cup brandy, dessert wine, or champagne (optional)
- 1 cup almonds; or use pecans, walnuts, hazelnuts, or macadamia nuts
- 3/4 cup pitted and packed dried fruit, like dates, raisins, dried cherries, figs, or apricots
- 4 ounces good-quality bittersweet chocolate, melted (optional)

1 Put the fruit in a bowl. Sprinkle with the sugar, and the liquor if you're using it. Toss gently to coat, and refrigerate while you prepare the crust.

2 Put the nuts in a food processor and pulse until ground, being careful not to overprocess. Transfer to a bowl, then put the dried fruit in the food processor along with a teaspoon or so of water. Pulse until finely chopped and sticky enough to adhere to the nuts (some fruit will require more water than others). Use your hands or a rubber spatula to combine the nuts and fruit in the bowl until they become a "dough." (At this point, you can form the mixture into a disk, wrap it in plastic, and refrigerate or freeze until about 30 minutes before you're ready to use it; defrost if necessary and proceed with the recipe.)

3 Divide the dough into 8 pieces and press into 3-inch round disks on a piece of wax paper or parchment; they should be about 1/4 inch thick. Brush each disk with the melted chocolate in a thin, even layer if you like, and top with fresh fruit and any accumulated juices. Serve immediately.

Chocolate Fondue with Fresh Fruit

Makes: **8 servings**   Time: **20 minutes**

Fondue is back, with good reason. It's just plain fun, and it's ridiculously easy to pull together. The rules are simple: Use the best chocolate and the best fruit you can find. Strawberries, bananas, pineapple, kiwi, melon, peaches, apricots, and candied ginger all work, as long as the pieces are big enough to spear easily. If you want to experiment a bit, try adding a ½ teaspoon cinnamon and ancho chili powder for some spice, ¼ teaspoon of espresso powder, a tablespoon or two of your favorite liqueur, or the seeds scraped from the inside of a vanilla bean.

½ cup of heavy cream, plus more as needed

12 ounces of good-quality bittersweet chocolate, roughly chopped

Splash of cognac, rum, bourbon, or kirsch (optional)

Chopped nuts (optional)

About 2 pounds assorted fruit (trimmed, peeled, pitted, and sliced as needed), for dipping

1 Put the cream in a saucepan over medium heat and just bring to a boil. Remove from the heat and add the chocolate, stirring until smooth. Add the liqueur and chopped nuts if you're using them.

2 Transfer the chocolate mixture to a fondue pot with a low flame (or put in a heated bowl and eat within 15 minutes or so); if the fondue begins to feel a little stiff, add a little more cream and stir. Serve with fresh fruit for dipping.

Coconut and Nut Chews

Makes: **About 2 dozen** Time: **About 45 minutes**

Although these contain no butter, cream, egg yolks, or flour, they're fantastic, and you can vary them in many ways. Use walnuts, pecans, almonds, pistachios, or hazelnuts. Or replace up to a cup of the nuts or coconut with chocolate chunks or dried cherries. In fact, as long as you keep the total volume to 3 cups, you can make a "kitchen sink" cookie that includes a bit of everything.

1 cup sugar

1½ cups shredded unsweetened coconut

1½ cups chopped nuts

4 egg whites, lightly beaten until just foamy

1 teaspoon vanilla extract

Pinch salt

1 Heat the oven to 350°F. Combine all ingredients in a large bowl and mix well with a rubber spatula or your hands. Use a nonstick baking sheet, or line a baking sheet with parchment paper. Wet your hands and make small balls with the mixture, each measuring about 1 tablespoon, and put them on the prepared sheet about an inch apart.

2 Bake until light brown, about 15 minutes. Remove the baking sheet and cool on a rack for at least 30 minutes before eating. These keep well in a covered container for up to 3 days.

Nutty Oatmeal Cookies

Makes: **About 3 dozen regular-size cookies** Time: **About 1 hour**

Nuts, chopped dried apples or other fruit, coconut or a combination help oats produce a great cookie. Choose the first options throughout this recipe ingredient list, and you've got a surprisingly good vegan cookie; go for the last options for a more traditional take. And if you're feeling decadent, this recipe can handle an extra ½ cup of bits of semisweet or dark chocolate.

To make the spice flavors more prominent, bump up the amount of cinnamon to a teaspoon, and add ½ teaspoon of ground ginger and ¼ teaspoon each ground allspice and nutmeg. Sometimes I make these cookies bite-size, since they're fully loaded with good stuff and quite satisfying in small portions. But small cookies will bake much more quickly, so keep an eye on them. (Then tuck half away in the freezer for a rainy day.)

½ cup peanut oil or vegetable oil, or 8 tablespoons (1 stick) unsalted butter, softened

½ cup granulated sugar

½ cup packed brown sugar

¼ cup applesauce, or 2 eggs

1½ cups all-purpose flour

2 cups rolled oats (not instant)

1 cup (about 3 ounces) chopped dried apples, or other fruit

½ cup chopped walnuts or pecans

½ teaspoon ground cinnamon

Pinch salt

2 teaspoons baking powder

½ cup almond milk, rice milk, or oat milk, or cow's milk

½ teaspoon vanilla or almond extract

1 Heat the oven to 375°F. Use an electric mixer to cream the oil or butter and the sugars together; add the applesauce or eggs and beat until well blended.

2 Combine the flour, oats, fruit, nuts, cinnamon, salt, and baking powder in a bowl. Alternating with the milk, add the dry ingredients to the butter and sugar mixture by hand, a little at a time, stirring to blend. Stir in the vanilla.

3 Put teaspoon-size mounds of dough about 3 inches apart on ungreased baking sheets. Bake until lightly browned, 10 to 12 minutes. Cool for about 2 minutes on the sheets, then use a spatula to transfer the cookies to a rack to finish cooling. Store in a tightly covered container at room temperature for no more than a day or two.

Fruit Crisp, with Apple or Nearly Anything Else

Makes: **8 servings**   Time: **About an hour**

Crisps are the ultimate year-round dessert, since the only thing that changes is the fruit. In the fall, use apples and pears (maybe with a handful of fresh cranberries); in winter, try pineapple; in the spring, nothing beats rhubarb (you might want a little more sugar in this case), especially combined with strawberries; and in the summer, stone fruits and berries rule, alone or in combination.

- 6 cups sliced or chopped fruit (2 to 3 pounds), trimmed, peeled, pitted, and cored as necessary
- 1 teaspoon ground cinnamon (optional)
- ½ cup brown sugar, or to taste
- 5 tablespoons cold unsalted butter, cut into bits, plus butter for greasing the pan
- ½ cup rolled oats
- ½ cup whole wheat flour
- ¼ cup chopped walnuts or pecans
- ¼ cup shredded unsweetened coconut (optional)
- Pinch salt

1 Heat the oven to 400°F. In a bowl, gently toss the fruit with half the cinnamon (if you're using it) and spread it in a lightly buttered 8-inch square or 9-inch round baking pan.

2 Combine all the other ingredients, including the remaining cinnamon, in the container of a food processor. Pulse a few times, then process a few seconds more until everything is well incorporated but not uniform; it should look crumbly. To mix the ingredients by hand, soften the butter slightly, toss together the dry ingredients, then work in the butter with your fingertips, a pastry blender, or a fork.

3 Crumble the topping over the fruit and bake 30 to 40 minutes, until the topping is browned and the fruit is tender and bubbling. Serve hot, warm, or at room temperature.

Roasted Fruit, Sweet or Savory: Without a crust this recipe becomes an easy fruit compote. Omit everything in the recipe, except the fruit, 2 tablespoons of the butter, and the salt. Then melt the butter along with the salt and whatever seasonings you choose. To go sweet, add some or all of the brown sugar if you like, and warm spices like the cinnamon; or use ½ teaspoon ground cardamom, nutmeg, or cloves; or tuck a split vanilla bean into the mixture. To take the fruit in a more savory direction, skip the sugar, increase the salt, and try ground cumin, cayenne, caraway, or even a pinch of saffron. Toss the fruit with the butter mixture, spread on the prepared pan, and cook as above, stirring once or twice.

Brown Rice Pudding

Makes: **4 servings** Time: **About 2½ hours, largely unattended**

To make rich pudding with brown rice, you have to bust the grains up a bit in the food processor so they will release enough starch to thicken the mixture. This nondairy recipe is based on coconut milk, but go ahead and use cow's milk for some or all of the liquid. And if you want a thicker pudding with more rice, veer toward the high end of the range listed below.

- ⅓ to ½ cup long-grain brown rice
- 3½ cups coconut milk (two 14-ounce cans)
- ½ cup brown sugar
- Small pinch salt
- Cinnamon stick, a few cardamom pods, a split vanilla bean, a pinch of saffron, or other flavoring (optional)

1 Heat the oven to 300°F. Put the rice in a food processor and pulse a few times to break the grains up a bit and scratch up their hulls; don't overprocess, or you'll pulverize them.

2 Put all the ingredients in a 2-quart ovenproof pot or Dutch oven. Stir a couple of times, and put in the oven, uncovered. Cook for 45 minutes, then stir. Cook for 45 minutes more, and stir again; at this point the milk will have darkened a bit and should be bubbling, and the rice will have begun to swell.

3 Cook for 30 minutes more. The milk will be even darker this time, and the pudding will start to look more like rice than milk. It's almost done. Return the mixture to the oven and check every 10 minutes, stirring gently each time you check.

4 It might—but probably won't—take as long as 30 minutes more for the pudding to be ready. You must trust your instincts and remove the custard from the oven when it is still soupy; it will thicken a lot as it cools. (If you overcook the pudding, it will become fairly hard, though it's still quite good to eat.) Remove the spices, if using; and serve the pudding warm, at room temperature, or cold, alone or with your favorite topping.

Some Ideas for Flavoring Brown Rice Pudding

Add 1 cup of shredded unsweetened coconut to the pot at the beginning.

About halfway through cooking, add 1/4 cup or more raisins, dried berries, or chopped dates, figs, or other dried fruit.

Stir 1 cup of chopped mango, papaya, or pineapple into the mix about halfway through the cooking.

Instead of the spices, add 1 teaspoon of vanilla extract, orange blossom water, or rose water at the end of cooking.

Add 1 teaspoon minced lemon or orange zest instead of the spices.

Garnish with a sprinkling of toasted coconut, sliced almonds, or other chopped nuts.

Chocolate Semolina Pudding with Raspberry Puree

Makes: **9 to 12 servings** Time: **1 hour**

Somewhere between a cake and pudding, this lovely dessert is served warm, with a simple raspberry puree that balances its richness. Other fruits that work well here include stone fruits, but (except for cherries) they have to be peeled first (page 149). Figure on about a pound of fruit for just over a cup of puree.

- 4 tablespoons (1/2 stick) butter, plus butter for the pan
- 1/4 cup cocoa powder
- 1/3 cup (2 ounces) semisweet or bittersweet chocolate, chopped
- 1 cup whole-milk yogurt
- 3/4 cup sugar
- 1 cup semolina
- 1/2 teaspoon baking soda
- 1 teaspoon vanilla extract
- 1 pound fresh raspberries
- Sugar (optional)
- Freshly squeezed lemon juice (optional)

1 Preheat the oven to 375°F. Grease an 8- or 9-inch square baking pan. Put the butter in a skillet over medium-high heat. When the foam subsides, add cocoa powder and semisweet or bittersweet chocolate and stir until smooth. Remove from the heat.

2 Beat the yogurt and sugar together in a large bowl. Add the butter and chocolate, the semolina, the baking soda, and the vanilla; beat until thoroughly blended. Spread the batter in the prepared pan. Bake until the pudding is lightly browned, about 30 minutes.

3 Meanwhile, puree the raspberries in a blender or food processor. Depending on how flavorful they are, you may want to add a tablespoon of sugar or a squeeze of lemon juice to the mixture, but taste first to see if either is necessary. Then strain the puree, stirring and pressing the mixture through a sieve with a rubber spatula to leave any seeds behind; be sure to get all the puree from the underside of the strainer.

4 When the pudding is done, let it rest for a few minutes, then cut it into squares or rectangles and serve warm, on some of the puree, with a few whole berries on top.

Other Dishes in the Book You Can Eat for Dessert

A Very Flexible Fruit Salad (page 162)

Fruit Smoothies (page 163)

Breakfast Bread Pudding (page 174)

Warm Nuts and Fruit (page 218)

Brown-Bag Popcorn (page 235)

Fruit and Cereal Bites (page 232)

Super-Simple Sorbet

Makes: **About 2 cups** Time: **10 minutes, with already frozen fruit**

No ice cream maker? No problem. This sorbet uses a food processor to turn frozen fruit and a bit of something creamy into a delicious frozen dessert. Store-bought frozen fruit makes this a snap all year round; in summer, just wash and freeze whatever you bring home from the market, and a couple hours later you're ready to go. (All stone fruit works beautifully; peel it first—see page 149.)

Some ideas: Honeydew and cantaloupe, especially with a good squeeze of lemon or lime; bananas (use lemon juice to help keep them from turning brown); cucumbers with a bit of jalapeño chile; berries. For chocolate cherry sorbet, skip the sugar, add 4 ounces of melted bittersweet chocolate, and use 12 ounces of frozen cherries.

As long as you keep the total volume of solid ingredients to about a pound, the combinations are endless.

1 pound frozen strawberries or other fruit

1/2 cup yogurt, crème fraîche, or silken tofu

1/4 cup sugar, more or less

Water as needed

Put all the ingredients except the water into a food processor, and process until pureed and creamy, stopping to scrape down the sides of the bowl as needed. If the fruit doesn't break down completely, gradually add some water through the feed tube a tablespoon or two at a time, being careful not to overprocess the sorbet into liquid. Serve immediately or freeze. To serve later, just allow 10 to 15 minutes for the sorbet to soften at room temperature.

BOOKS

Anderson, E. N. *Everyone Eats: Understanding Food and Culture.* New York: New York University Press, 2005.

Brown, Lester R. *Outgrowing the Earth: The Food Security Challenge in an Age of Falling Water Tables and Rising Temperatures.* New York: W.W. Norton, 2004.

———. *Plan B 2.0: Rescuing a Planet Under Stress and a Civilization in Trouble.* New York: W.W. Norton, 2006.

Fuhrman, Joel. *Eat to Live: The Revolutionary Formula for Fast and Sustained Weight Loss.* New York: Little, Brown and Company, 2003.

Gaesser, Glenn A. *Big Fat Lies: The Truth About Your Weight and Your Health,* Updated Edition. Carlsbad, CA: Gurze Books, 2002.

Gaesser, Glenn A., and Karin Kratina. *It's the Calories Not the Carbs.* Victoria, BC, Canada: Trafford Publishing, 2006.

Gaesser, Glenn A., and Karla Dougherty. *The Spark: The Revolutionary New Plan to Get Fit and Lose Weight 10 Minutes at a Time.* New York: Fireside, 2002.

Jackson, Wes. *New Roots for Agriculture.* Lincoln, NE: University of Nebraska Press, 1980.

Lang, Tim, and Michael Heasman. *Food Wars: Public Health and the Battle for Mouths, Minds and Markets.* London: Earthscan, 2004.

Lappé, Frances Moore. *Diet for a Small Planet.* New York: Ballantine Books, 1982.

Lappé, Frances Moore, and Anna Lappé. *Hope's Edge: The Next Diet for a Small Planet.* New York: Jeremy P. Tarcher/Putnam, 2003.

Lappé, Anna, and Bryant Terry. *Grub: Ideas for an Urban Organic Kitchen.* New York: Jeremy P. Tarcher/Penguin, 2006.

Mason, Jim. *An Unnatural Order: Why We Are Destroying the Planet and Each Other.* New York: Continuum, 1997.

Nabhan, Gary Paul. *Coming Home to Eat: The Pleasures and Politics of Local Foods.* New York: W.W. Norton, 2002.

Nestle, Marion. *Food Politics: How the Food Industry Influences Nutrition and Health.* Berkeley: University of California Press, 2002.

———. *Safe Food: Bacteria, Biotechnology, and Bioterrorism.* Berkeley: University of California Press, 2003.

———. *What to Eat.* New York: North Point Press, 2007.

Nestle, Marion, and L. Beth Dixon. *Taking Sides: Clashing Views on Controversial Issues in Food and Nutrition.* Guilford, CT: McGraw-Hill/Dushkin, 2003.

Pimentel, David, and Marcia Pimentel. *Food, Energy, and Society:* Third Edition. Florida: CRC Press, 2008.

Reisner, Mark. *Cadillac Desert: The American West and Its Disappearing Water,* Revised Edition. New York: Penguin Books, 1993.

Rifkin, Eric, and Edward Bouwer. *The Illusion of Certainty: Health Benefits and Risks.* New York: Springer, 2007.

Robb, Jay. *Be Still, Be Slim: The Peaceful Way to Lose Weight—and Keep It Off.* Carlsbad, CA: Loving Health Publications, 2007.

Robbins, John. *Diet for a New America.* Tiburon, CA: H.J. Kramer, 1998.

———. *The Food Revolution: How Your Diet Can Help Save Your Life and Our World.* Berkeley: Conari Press, 2001.

Simon, Michele. *Appetite for Profit: How the Food Industry Undermines Our Health and How to Fight Back.* New York: Nation Books, 2006.

Singer, Peter, and Jim Mason. *The Ethics of What We Eat: Why Our Food Choices Matter.* New York: Rodale Books, 2007.

Steinman, David. *Diet for a Poisoned Planet: How to Choose Safe Foods for You and Your Family—The Twenty-first Century Edition.* New York: Running Press, 2006.

Swain, Brian Kenneth. *World Hunger.* Lincoln, NE: iUniverse, 2007.

Taubes, Gary. *Good Calories, Bad Calories: Challenging the Conventional Wisdom on Diet, Weight Control and Disease*. New York: Knopf, 2007.

Tudge, Colin. *So Shall We Reap: What's Gone Wrong with the World's Food—and How to Fix it.* New York: Penguin Books Ltd., 2004.

Volek, Jeff, and Adam Campbell. *Men's Health TNT Diet: The Explosive New Plan to Blast Fat, Build Muscle, and Get Healthy in 12 Weeks.* New York: Rodale Books, 2007.

Watson, James L., and Melissa L. Caldwell, eds. The *Cultural Politics of Food and Eating.* Oxford: Blackwell, 2005.

Westcott, Wayne, and Gary Reinl. *Get Stronger, Feel Younger: The Cardio and Diet-Free Plan to Firm Up and Lose Fat.* New York: Rodale Books, 2007.

Willcox, Bradley J., et al. *The Okinawa Diet Plan.* New York: Three Rivers Press, 2004.

Willett, Walter C., and Patrick J. Skerrett. *Eat, Drink, and Be Healthy: The Harvard Medical School Guide to Healthy Eating.* New York: Free Press, 2005.

JOURNALS, MAGAZINES, ONLINE RESOURCES, AND OTHER PUBLICATIONS

Activia Yogurt Web Site. http://www.activia.us.com/pdf/Act_scientific_summary.pdf?v1.

American Council for an Energy-Efficient Economy. "Introduction: Setting Priorities, Reducing Your Impact." *Consumer Guide to Home Energy Savings: Condensed Online Version*, 2007. http://www.aceee.org/consumerguide/intro.htm.

American Council for an Energy-Efficient Economy. "Cooking." *Consumer Guide to Home Energy Savings: Condensed Online Version*, 2007. http://www.aceee.org/consumerguide/cooking.htm.

American Heart Association. *Heart Disease and Stroke Statistics: 2005 Update*. http://www.americanheart.org/downloadable/heart/1105390918119HDSStats2005Update.pdf.

American Heart Association. "Fiber." http://www.americanheart.org/presenter.jhtml?identifier=4574.

Anderson, Douglas. "Obesity, Jellybeans and Soda." *Dynamic Chiropractic*, March 26, 2001. http://findarticles.com/p/articles/mi_qa3987/is_200103/ai_n8930669.

Associated Press. "Study: Atkins Diet Good for Cholesterol." *USA Today*, November 18, 2002. http://www.usatoday.com/news/health/2002-11-18-atkins_x.htm.

Avila, Jim, and Reynolds Holding. "Dannon Yogurt Faces Lawsuit Over False Advertising." *ABC News*, January 25, 2008. http://www.abcnews.go.com/thelaw/Story?id=4188726&page=1

Becker, Geoffrey S. *CRS Report for Congress: Farm and Food Support Under USDA's Section 32 Program, 2007*. http://www.nationalaglawcenter.org/assets/crs/RS20235.pdf.

The Biology Project. "The Chemistry of Amino Acids." http://www.biology.arizona.edu/biochemistry/problem_sets/aa/aa.html.

Block, Gladys. "Foods Contributing to Energy Intake in the US: Data from NHANES III and NHANES 1999–2000." *Journal of Food Composition and Analysis*, 17 (2004): 439–447.

Blue Water Network. http://www.bluewaternetwork.org/faqs.shtml.

Bray, George A., et al. "Consumption of High-Fructose Corn Syrup in Beverages May Play a Role in the Epidemic of Obesity." *American Journal of Clinical Nutrition*, 79 (2004): 537–543. http://www.ajcn.org/cgi/content/full/79/4/537.

The Canadian Press. "Eat Less Meat, Reduce Global Heat, Says Study." *CBC.ca*, September 13, 2007. http://www.cbc.ca/health/story/2007/09/13/meat-study.html.

Carey, Benedict. "Antidepressant Studies Unpublished." *New York Times*, January 17, 2008. http://www.nytimes.com/2008/01/17/health/17depress.html?_r=1&oref=slogin.

Carlsson-Kanyama, Annika, and Mireille Faist. "Energy Use in the Food Section: A Data Survey." AFR Report 219, 2000. http://www.infra.kth.se/fms/pdf/energyuse.pdf.

Cascio, Jamais. "The Cheeseburger Footprint." *Open the Future*. http://openthefuture.com/cheeseburger_CF.html.

CDC (Centers for Disease Control and Prevention). "Chronic Disease Prevention: Economic and Health Burden of Chronic Disease." http://www.cdc.gov/nccdphp/press/#3.

———. "Chronic Disease Prevention: Preventing Heart Disease and Stroke." http://www.cdc.gov/nccdphp/publications/factsheets/Prevention/cvh.htm.

———. "Overweight and Obesity: Introduction." http://www.cdc.gov/nccdphp/dnpa/obesity/.

CDC Press Release. "Number of Cancer Survivors Growing According to New Report." June 24, 2004. http://www.cdc.gov/od/oc/media/pressrel/r040624.htm.

Center for Science in the Public Interest. "Dietary Guidelines Committee Criticized." August 19, 2003. http://cspinet.org/new/200308191.html.

———. *Dispensing Junk: How School Vending Undermines Efforts to Feed Children Well*, 2004. http://cspinet.org/new/pdf/dispensing_junk.pdf.

Central Intelligence Agency. *The World Fact Book*. "Rank Order: Life Expectancy at Birth." https://www.cia.gov/library/publications/the-world-factbook/rankorder/2102rank.html.

"Checkoff Leverages Funding for Prostate Cancer Study." *Southwest Farm Press*, June 5, 2003. http://southwestfarmpress.com/mag/farming_checkoff_leverages_funding/.

Child and Family Research Institute Press Release. "Too Much Sugar Turns Off the Gene That Controls the Effects of Sex Steroids." November 8, 2007. http://www.eurekalert.org/pub_releases/2007-11/cfr-tms110907.php.

Compassion in World Farming Report. *Global Warming: Climate Change and Farm Animal Welfare*, 2007. http://www.ciwf.org/publications/reports/global-warning.pdf.

Cordain, Loren, et al. "Origins and Evolution of the Western Diet: Health Implications for the 21st Century." *American Society for Clinical Nutrition*, 81 (2005): 341–354. http://www.ajcn.org/cgi/content/full/81/2/341.

Corn Refiner's Association. "Statistics: U.S. Per Capita Sweetener Deliveries for Food and Beverage Use." http://www.corn.org/percapsw.htm.

———. "U.S. Corn Supply and Disappearance." http://www.corn.org/suppdisp.htm.

Davis, Carole, and Etta Saltos. "Dietary Recommendations and How They Have Changed Over Time." *America's Eating Habits: Changes and Consequences.* USDA Economic Research Service, 1999: 33–50. http://www.ers.usda.gov/publications/aib750/aib750b.pdf.

Delgado, Christopher L. "Rising Consumption of Meat and Milk in Developing Countries Has Created a New Food Revolution." *Journal of Nutrition,* 133 (2003): 3907S–3910S. http://jn.nutrition.org/cgi/contentfull/133/11/3907S.

"Drug Company Funding of Drug Trials Greatly Influences Outcome." *Science Daily,* June 5, 2007. http://www.sciencedaily.com/releases/2007/06/070604205602.htm.

Edelson, Ed. "Walnuts May Beat Olive Oil for Heart Health." *Health-Day News.* http://www.memorialhermann.org/healthyliving/nutrition/sept07walnuts.htm.

Elliott, Stuart. "The Media Business: Advertising; Disney, and McDonald's as Double Feature." *New York Times,* May 24, 1996. http://query.nytimes.com/gst/fullpage.html?res=9901E4D61039F937A15756C0A960958260&sec=&spon=&pagewanted=all.

Eshel, Gidon, and Pamela A. Martin. "Diet, Energy, and Global Warming." *Earth Interactions,* 10 (2006): 1–17. http://geosci.uchicago.edu/~gidon/papers/nutri/nutriEI.pdf

Esselstyn, C.B., Jr., et al. "A Strategy to Arrest and Reverse Coronary Artery Disease: A 5-Year Longitudinal Study of a Single Physician's Practice." *Journal of Family Practice,* 41 (1995): 560–568. http://www.ncbi.nlm.nih.gov/pubmed/7500065.

Fanelli, Danielle. "Meat Is Murder on the Environment." *New Scientist,* July 18, 2007. http://environment.newscientist.com/article/mg19526134.500-meat-is-murder-on-the-environment.html.

FAO Newsroom. "Farm Animal Diversity Under Threat." June 14, 2007. http://www.fao.org/newsroom/en/news/2007/1000598/index.html.

———. "Livestock a Major Threat to Environment." November 29, 2006. http://www.fao.org/newsroom/ennews/2006/1000448/index.html.

FAO, IFAD, and WFP. "Reducing Poverty and Hunger: The Critical Role of Financing for Food, Agriculture, and Rural Development." *International Conference on Financing for Development,* March 2002. ftp://ftp.fao.org/docrep/fao/003/y6265e/y6265e.pdf.

FDA. "Making Sense of the Cholesterol Controversy." http://69.20.19.211/bbs/topics/CONSUMER/CON00052.html.

FDA, Center for Food Safety and Applied Nutrition. "Claims That Can Be Made for Conventional Foods and Dietary Supplements." September 2003. http://www.cfsan.fda.gov/~dms/hclaims.html.

———. "A Food Labeling Guide: XI. Appendix C: Health Claims." April 2008. http://www.cfsan.fda.gov/~dms/2lg-xc.html.

"Fed Up: America's Killer Diet." Transcript of Interview by Dr. Sanjay Gupta for CNN, aired September 22, 2007. http://transcripts.cnn.com/TRANSCRIPTS/0709/22/siu.03.html.

The Franklin Institute. http://www.fi.edu/learn/brain/fats.html.

Gardner, Amanda. "Atkins Diet Can Raise Heart Risks." *U.S. Department of Health & Human Services.* http://www.healthfinder.gov/news/newsstory.asp?docID=609782.

Gross, Lee S., et al. "Increased Consumption of Refined Carbohydrates and the Epidemic of Type 2 Diabetes in the United States: An Ecologic Assessment." *American Journal of Clinical Nutrition*, 79 (2004): 774–779. http://www.ajcn.org/cgi/content/abstract/79/5/774?maxtoshow=&HITS=10&hits=10&RESULTFORMAT=&author1=Gross&searchid=1084292919397_7392&stored_search=&FIRSTINDEX=0&sortspec=relevance&volume=79&firstpage=774&journalcode=ajcn.

Halweil, Brian. "The Bioethics of Barbecue: Environmental Consequences of Eating Massive Amounts of Meat." *MSNBC,* June 30, 1998. http://www.cybergeo.com/wellness/bioethics.html.

Halweil, Brian, and Danielle Nierenberg. "Meat and Seafood: The Most Costly Ingredients in the Global Diet." *State of the World*. WorldWatch, 2008, 61–228. http://www.worldwatch.org/node/5561.

Harper, A.E. "Origin of Recommended Dietary Allowances: An Historic Overview." *American Journal of Clinical Nutrition*, 41 (1985): 140–148. http://www.ajcn.org/cgi/reprint/41/1/140.pdf.

Harvard School of Public Health. "The Nutrition Source: Carbohydrates." http://www.hsph.harvard.edu/nutritionsource/what-should-you-eat/carbohydrates/.

———. "The Nutrition Source: Fats and Cholesterol." http://www.hsph.harvard.edu/nutritionsource/what-should-you-eat/fats-and-cholesterol/.

———. "The Nutrition Source: Fiber." http://www.hsph.harvard.edu/nutritionsource/what-should-you-eat/fiber/.

———. "The Nutrition Source: Health Gains from Whole Grains." http://www.hsph.harvard.edu/nutritionsource/what-should-you-eat/health-gains-from-whole-grains/.

———. "The Nutrition Source: Low-Fat Diet Not a Cure-All." http://www.hsph.harvard.edu/nutritionsource/low_fat.html.

———. "The Nutrition Source: Protein." http://www.hsph.harvard.edu/nutritionsource/Printer%20Friendly/Protein.pdf.

———. "The Nutrition Source: Type 2 Diabetes." http://www.hsph.harvard .edu/nutritionsource/more/type-2-diabetes/.

Harvard University Health Services. "Protein Nutrition." http://huhs.harvard .edu/Resources/HealthInformationByTopic/Nutrition/ProteinNutrition.aspx.

Hermann, Janice R. "Protein and the Body." *Oklahoma Cooperative Extension Fact Sheets.* http://pods.dasnr.okstate.edu/docushare/dsweb/Get/Document-2473/T-3163web.pdf.

Hill, Holly. *Food Miles: Background and Marketing.* A Publication of ATTRA—National Sustainable Agriculture Information Service, 2008. http:// attra.ncat.org/attra-pub/PDF/foodmiles.pdf.

Horovitz, Bruce. "Yogurt Cultivating Unprecedented Popularity." *USA Today,* January 23, 2006. http://www.usatoday.com/money/industries/food/2006-01-23-yogurt-usat_x.htm.

Humane Society of the United States. *An HSUS Report: The Welfare of Animals in the Meat, Egg, and Dairy Industries,* 2006. http://www.hsus.org/ farm/resources/research/welfare/welfare_overview.html.

Institute for Agriculture and Trade Policy. *Antibiotics and Fish Farming,* 2002. http://www.healthobservatory.org/library.cfm?refID=70160.

Jenkins, David J.A., et al. "Assessment of the Longer-Term Effects of a Dietary Portfolio of Cholesterol-Lowering Foods in Hypercholesterolemia." *American Journal of Clinical Nutrition*, 83 (2006): 582–591. http://www.ajcn.org/cgi/ reprint/83/3/582.pdf.

Kaufman, Marc. "Antibiotics in Animal Feed: A Growing Public Health Hazard." *Organic Consumer's Association,* March 17, 2000. http://www.organic consumers.org/Toxic/animalfeed.cfm.

Klein, Benjamin, and Joshua D. Wright. "The Economics of Slotting Contracts." *Journal of Law and Economics,* 2007. http://www.ftc.gov/os/sectiontwohear ings/docs/Slotting_paper_110806.pdf.

Knickerbocker, Brad. "Human's Beef with Livestock: A Warmer Planet." *Christian Science Monitor,* February 20, 2007. http://www.csmonitor.com/ 2007/0220/p03s01-ussc.html.

Leibman, Bonnie. "Cereal Sinners: Healthy Marketing Ploy by Cereal Companies." *Nutrition Action Healthletter,* March 1999. http://findarticles.com/p/ articles/mi_m0813/is_2_26/ai_53996209.

Lesser, Lenard I., et al. "Relationship Between Funding Source and Conclusion Among Nutrition-Related Scientific Articles." *PLOS Medicine.* http://medicine .plosjournals.org/perlserv/?request=get-document&doi=10.1371%2Fjournal .pmed.0040005.

Lichtenstein, Alice H., and Linda Van Horn. "AHA Science Advisory: Very Low Fat Diets." *Circulation*, 98 (1998): 935–939. http://circ.ahajournals.org/cgi/content/full/98/9/935http://circ.ahajournals.org/cgi/content/full/98/9/935.

Light, Luise. "A Fatally Flawed Food Guide." *Conscious Choice,* November 2004. http://www.consciouschoice.com/2004/cc1711/wh_lead1711.html.

Lochhead, Carolyn. "Huge Farm Bill Offers More of Same for Agribusiness." *SF Gate,* July 26, 2007. http://www.sfgate.info/cgi-bin/article.cgi?f=/c/a/2007/07/26/MNG9AR6V6S1.DTL.

McDonald's USA Nutritional Facts for Popular Menu Items. http://app.mcdonalds.com/countries/usa/food/nutrition/categories/nutritionfacts.pdf.

McMichael, Anthony J., et al. "Food, Livestock Production, Energy, Climate Change, and Health." *The Lancet.com,* September 13, 2007. http://www.eurekalert.org/images/release_graphics/pdf/EH5.pdf.

Medline Plus, Medical Encyclopedia. http://www.nlm.nih.gov/medlineplus/ency/article/002467.htm#Definition.

Molotsky, Irvin. "Animal Antibiotics Tied to Illnesses in Humans." *New York Times,* February 22, 1987. http://query.nytimes.com/gst/fullpage.html?res=9BODE4D91531F931A15751COA961948260&sec=&spon=&pagewanted=all.

National Agricultural Law Center. http://www.nationalaglawcenter.org/assets/farmbills/glossary.html#.

National Corn Growers Association. http://www.ncga.com/03world/main/production.htm.

NIH (National Institutes of Health). National Cancer Institute. *Cancer Trends Progress Report: 2007 Update*. http://progressreport.cancer.gov/highlights.asp.

———. National Cancer Institute. "Low-Fat Diet May Reduce Risk of Breast Cancer Relapse." http://www.cancer.gov/clinicaltrials/results/low-fat-diet0505.

———. National Diabetes Information Clearinghouse. "Diabetes Statistics, 2007." http://diabetes.niddk.nih.gov/dm/pubs/statistics/index.htm#allages.

———. National Digestive Disease Information Clearinghouse. "Lactose Intolerance." http://digestive.niddk.nih.gov/ddiseases/pubs/lactose intolerance/.

Northwestern University. "Nutrition Fact Sheet: Dietary Fiber." http://www.feinberg.northwestern.edu/nutrition/factsheets/fiber.html.

Ohlson, Kristin. "America's Appetite for Olive Oil Ripens." *Christian Science Monitor,* January 10, 2007. http://www.csmonitor.com/2007/0110/p13s01-lifo.html.

Ornish, Dean, et al. "Can Lifestyle Changes Reverse Coronary Heart Disease?

The Lifestyle Heart Trial." *The Lancet, 1990*. http://www.ncbi.nlm.nih.gov/pubmed/1973470?dopt=Abstract.

Physicians Committee for Responsible Medicine. *School Lunch Report Card*, 2003. http://www.pcrm.org/health/reports/schoollunch_report2003.html.

Pimentel, David, and Marcia Pimentel. "Sustainability of Meat-based and Plant-based Diets and the Environment." *American Journal of Clinical Nutrition*, 78 (2003): 660S–663S. http://www.ajcn.org/cgi/content/full/78/3/660S?ck=nck.

Pollan, Michael. "Power Steer." *New York Times Magazine*, March 31, 2002. http://www.michaelpollan.com/article.php?id=14.

———. "Unhappy Meals." *New York Times Magazine*, January 28, 2007. http://www.nytimes.com/2007/01/28/magazine/28nutritionism.t.html?_r=1&oref=slogin.

Popcorn Board. http://www.popcorn.org/about/checkoffs.cfm.

"Positive Results More Likely From Industry-Funded Breast Cancer Trials." *Science Daily*, February 26, 2007. http://www.sciencedaily.com/releases/2007/02/070226095116. htm.

Preidt, Robert. "U.S. Heart Disease Death Rates Falling." *Health Day*, December 18, 2007. http://www.nlm.nih.gov/medlineplus/news/fullstory_59053.html.

Quinion, Michael. *World Wide Words*. http://www.worldwidewords.org/turnsofphrase/tp-gl02.htm.

Radak, Tim. "Disaster by Design: Confusing New Food Pyramid Misleads Consumers About Healthy Eating." Physicians Committee for Responsible Medicine, News and Media Center, May 2005. http://www.pcrm.org/news/commentary0505.html.

Raloff, Janet. "Hormones: Here's the Beef." *Science News*, January 5, 2002. http://www.sciencenews.org/articles/20020105/bob13.asp.

"The Rise and Rise of Acai in the U.S." *New Nutrition Business*, April 2007. http://www.nufruits.com/documents/quotes.pdf.

Ritchie, Mark, and Kevin Ristau. *Crisis By Design: A Brief Review of U.S. Farm Policy*. League of Rural Voters Education Project, 1987. http://www.iatp.org/iatp/publications.cfm?accountID=258&refID=48644.

Robinson, Thomas N., et al. "Effects of Fast Food Branding on Young Children's Taste Preferences." *Archives of Pediatrics & Adolescent Medicine*, 161 (2007): 792–797. http://archpedi.ama-assn.org/cgi/content/full/161/8/792.

Salt Institute, a Statement by the USDA. "Facts About Nutrition and Health." February 2000. http://www.saltinstitute.org/usda-facts.html.

Severson, Kim. "Sugar Coated." *SFGate.com*, February 18, 2004. http://www.sfgate.com/cgi-bin/article.cgi?f=/chronicle/archive/2004/02/18/FDGS24VKMH1.DTL.

Smith, Aaron. "Big Pharma's Drug Wish List for 2007." *CNNMoney.com*, December 21, 2006. http://money.cnn.com/2006/12/21/news/companies/topfivedrugs/index.htm.

"Some Studies More Likely to Show Bias in Favor of Funding Body Than Others." *Science Daily*, November 18, 2007. http://www.sciencedaily.com/releases/2007/11/071116094755.htm.

"Soybean Checkoff Promotes U.S. Appetite for Soyfoods." *Southwest Farm Press*, June 5, 2003. http://southwestfarmpress.com/mag/farming_soybean_checkoff_promotes/.

Soy Info Center. "A Special Report on the History of Soy Oil, Soybean Meal, and Modern Soy Protein Products." A Chapter from the Unpublished Manuscript, *History of Soybeans and Soyfoods: 1100 B.C. to the 1980s* by William Shurtleff and Akiko Aoyagi, 2007. http://www.soyinfocenter.com/HSS/margarine3.php.

Stanton, T.L., and S.B LeValley. "Feed Composition for Cattle and Sheep." Colorado State University Extension. http://www.ext.colostate.edu/PUBS/livestk/01615.html.

Steinfeld, Henning, et al. *Livestock's Long Shadow: Environmental Issues and Options*. FAO, Rome, 2006. http://www.virtualcentre.org/en/library/key_pub/longshad/A0701E00.htm.

Stonyfield Farm Web site. http://www.stonyfield.com/OurProducts/.

"The 2002 Farm Bill: Implications for International Trade." Transcript of Briefing by J.B. Penn, Under Secretary of Agriculture for Farm and Foreign Agricultural Services, for U.S. Department of State, May 22, 2002. http://fpc.state.gov/fpc/10408.htm.

United Nations Press Release. "World Population to Increase by 2.6 Billion Over Next 45 Years." 2005. http://www.un.org/News/Press/docs/2005/pop918.doc.htm.

USDA (United States Department of Agriculture). "An Economic Overview of Horticultural Products in the United States." http://www.fas.usda.gov/htp/Presentations/2004/An%20Economic%20Overview%20of%20HTP%20-%20(08-04).pdf.

———. "Appendix G-5: History of the Dietary Guidelines for Americans." *Nutrition and Your Health: Dietary Guidelines for Americans.* http://www.health.gov/dietaryguidelines/dga2005/report/HTML/G5_History.htm.

———. *Dietary Guidelines for Americans, 1995.* http://www.nal.usda.gov/fnic/dga/dga95/lowfat.html.

———. *Dietary Guidelines for Americans, 2005.* http://www.health.gov/dietaryguidelines/dga2005/document/html/chapter6.htm.

———. *FY 2009: Budget Summary and Performance Plan.* http://www.obpa.usda.gov/budsum/fy09budsum.pdf.

———. "Profiling Food Consumption in America." *Agricultural Fact Book 2001–2002.* http://www.usda.gov/factbook/chapter2.htm.

USDA Agricultural Marketing Service. "National Organic Program." http://www.ams.usda.gov/nop/Consumers/brochure.html.

USDA Cooperative State Research, Education, and Extension Service. http://www.csrees.usda.gov/qlinks/extension.html.

USDA Economic Research Service. *Food Security Assessment, 2007*. http://www.ers.usda.gov/Publications/GFA19/.

———. *Fruit and Tree Nuts Outlook*, 2005. http://www.ers.usda.gov/publications.

———. *Let's Eat Out: Americans Weigh Taste, Convenience, and Nutrition*, October 2006. http://www.ers.usda.gov/publications/eib19/eib19_report summary.pdf.

USDA Food and Nutrition Service. *National School Lunch Program*, July 2007. http://www.fns.usda.gov/cnd/lunch/aboutlunch/NSLPFactSheet.pdf

———. *School Nutrition Dietary Assessment Study-III: Summary of Findings*, 2007. http://www.fns.usda.gov/oane/menu/Published/CNP/FILES/SNDAIII-SummaryofFindings.pdf.

USDA Press Release. "Partner with My Pyramid: USDA Asks Industry to Step Up for Nutritious Choices." February 15, 2008. http://www.cnpp.usda.gov/Publications/News-Media/MyPyramidandIndustry-2-14-08.pdf.

United States Department of Health and Human Services. *Dietary Goals for the United States*, 1977. http://www.becomehealthynow.com/ebook print.php?id=157.

United States Environmental Protection Agency. "Major Crops Grown in the United States." http://www.epa.gov/oecaagct/ag101/cropmajor.html.

United States Social Security Administration. *The Fiscal Year 2008 Budget Press Release.* http://www.socialsecurity.gov/budget/2008bud.pdf.

Weber, Christopher L., and Scott H. Matthews. "Food Miles and the Relative Climate Impacts of Food Choices in the United States." *Environmental Science and Technology*, 42 (2008): 3508–3513. http://pubs.acs.org/cgi-bin/abstract.cgi/esthag/2008/42/i10/abs/es702969f.html

"White Bread Linked to Diabetes." *CBS News,* November 5, 2004. http://www.cbsnews.com/stories/2004/11/05/health/webmd/main653957.shtml.

Wolpert, Stuart. "Dieting Does Not Work, UCLA Researchers Report." *UCLA Newsroom,* July 9, 2008. http://www.newsroom.ucla.edu/portal/ucla/Dieting-Does-Not-Work-UCLA-Researchers-7832.aspx?RelNum=7832.

Yang, Sarah. "Nearly One-Third of the Calories in the U.S. Diet Come from Junk Food, Researcher Finds." *UC Berkeley News,* June 1, 2004. http://www.berkeley.edu/news/media/releases/2004/06/01_usdiet.shtml.

DATABASES

Dietary Reference Intakes. http://www.iom.edu/Object.File/Master/7/300/0.pdf.

FAOSTAT. Food and Agriculture Organization of the United Nations (FAO). Animal production online database FAOSTAT, 2006. http://faostat.fao.org/default.aspx.

US Census Bureau. "Infant Mortality Rates and Deaths, and Life Expectancy at Birth, by Sex." Table 010. http://www.census.gov/cgi-bin/ipc/idbagg.

USDA Agricultural Research Service, Nutrient Data Laboratory. http://www.nal.usda.gov/fnic/foodcomp/search/.

———. National Nutrient Database for Standard Reference, Release 20 (2007). http://www.ars.usda.gov/services/docs.htm?docid=8964.

———. "Research Project: Dietary Factors Early in Human Development, Health Consequences of Phytochemical Intake." http://www.ars.usda.gov/research/publications/htm?seq_no_115=214234.

USDA Economic Research Service. "Food Availability: Spreadsheets." http://www.ers.usda.gov/data/FoodConsumption/FoodavailSpreadsheets.htm#mtredsu.

———. "The 20th Century Transformation of U.S. Agriculture and Farm Policy." http://www.ers.usda.gov/publications/EIB3/EIB3.htm.

———. "Meat Price Spreads." http://www.ers.usda.gov/Data/meatpricespreads/.

———. "U.S. Beef and Cattle Industry: Background Statistics and Information." http://www.ers.usda.gov/news/BSECoverage.htm.

———. "Soybeans: Supply, Disappearance, and Price, U.S. 1980/81–2007/08." http://usda.mannlib.cornell.edu/ers/89002/Table03.xls.

USDA National Agriculture Statistics Service. "Livestock Slaughter 2006 Summary." March 2007. http://usda.mannlib.cornell.edu/usda/nass/LiveSlauSu//2000s/2007/LiveSlauSu-03-02-2007.pdf.

———. "Poultry Slaughter." December 2007. http://usda.mannlib.cornell.edu/usda/nass/PoulSlau//2000s/2007/PoulSlau-12-31-2007.pdf.

Common Measurements

VOLUME TO VOLUME

3 teaspoon = 1 tablespoon
4 tablespoons = ¼ cup
5 ⅓ tablespoons = ⅓ cup
4 ounces = ½ cup
8 ounces = 1 cup
1 cup = ½ pint

VOLUME TO WEIGHT

¼ cup liquid or fat = 2 ounces
½ cup liquid or fat = 4 ounces
1 cup liquid or fat = 8 ounces
2 cups liquid or fat = 1 pound
1 cup sugar = 7 ounces
1 cup flour = 5 ounces

Metric Equivalencies

LIQUID AND DRY MEASURES

CUSTOMARY	METRIC
¼ teaspoon	1.25 milliliters
½ teaspoon	2.5 milliliters
1 teaspoon	5 milliliters
1 tablespoon	15 milliliters
1 fluid ounce	30 milliliters
¼ cup	60 milliliters
⅓ cup	80 milliliters
½ cup	120 milliliters
1 cup	240 milliliters
1 pint (2 cups)	480 milliliters
1 quart (4 cups)	960 milliliters (0.96 liters)
1 gallon (4 quarts)	3.84 liters
1 ounce (by weight)	28 grams
¼ pound (4 ounces)	114 grams
1 pound (16 ounces)	454 grams
2.2 pounds	1 kilogram (1000 grams)

OVEN-TEMPERATURES

DESCRIPTION	°FAHRENHEIT	°CELSIUS
Cool	200	90
Very slow	250	120
Slow	300–325	150–160
Moderately slow	325–350	160–180
Moderate	350–375	180–190
Moderately hot	375–400	190–200
Hot	400–450	200–230
Very hot	450–500	230–260

ACKNOWLEDGMENTS

The three women to whom *Food Matters* is dedicated played huge roles in its production. Kerri Conan, my associate in the last several books I've written was, as usual, completely indispensable and a daily companion and co-conspirator. The idea for the book came to us while working with Suzanne Lenzer on a couple of smaller but seminal projects, and the three of us hatched it, developed it and, to a large extent, lived it. In a way, the traditional research, though long and grueling, was the easy part; developing a Food Matters diet—what we call a sane way of eating—was something we had to create.

In designing the book, Kelly Doe reshaped not only the way we thought it should look but the way it was organized. If it was not a completely different entity after she got her hands on it, it was certainly an improved one.

Sid Baker, my personal doctor for going on 30 years, was in part responsible for setting these wheels in motion by suggesting that I become a vegan despite knowing how ridiculous that suggestion was. The compromise between what Sid insisted and what was possible is essentially the Food Matters diet.

Sydny Miner, our editor, patiently and calmly stood by as we retooled many aspects of the manuscript and its design. This is as close to a breaking news book as I'll ever write, and right down to the deadline there were changes I thought worth making; Syd saw that and tolerated it as long as she could. Behind her stands an entire team at Simon & Schuster that has been a complete joy to work with, starting with publisher David Rosenthal (whose enthusiasm for the concept was immediate and intense), and continuing with Michelle Rorke, Nancy Singer, Deirdre Amthor and Larry Pekarek; Mara Lurie and Susan Gamer; Aileen Boyle and Deborah Darrock, Jackie Seow, Victoria Meyer and Alexis Welby.

One could go back to Hippocrates to thank for inspiration here, because if your diet did you no harm you'd be way ahead of the game compared to most Americans. But I think it's more helpful to thank those who are living, and there are (fortunately), some important heroes in the world of politics, nutrition, and food among us: Marion Nestle (whom I first heard utter the phrase "Food Matters," when she was discussing possible titles for her invaluable book, ultimately called *Food Politics*); Michael Pollan; Peter Singer; Wendell Berry; Frances Moore Lappé; Lester Brown; and Joel Fuhrman; I hope that I've honored their work by building on it. It may seem odd to thank what amount to faceless entities, but we could not have done this book without the help of the CDC, the UN FAO, and even the USDA.

From the time I began writing about food, Chris Kimball, Trish Hall, Rick Flaste, Kerri Conan (again), and always—always—Angela Miller, have encouraged me to take the subject of food more seriously than was common as did, for a relatively brief but important (and long ago!) time, Scott Mowbray. In recent years, I've gotten similar support from Sam Sifton, Pete Wells, Nick Fox, Chris Conway, and Mike Hawley.

Finally, for their personal support in a year that was even more whirlwind than others, I'd like to thank John Willoughby, my beloved parents and children, Gertrude, Murray, Kate, and Emma, and my wife Kelly.

New York
October 2008

Page numbers in *italics* refer to charts. Recipe index follows on page 319.